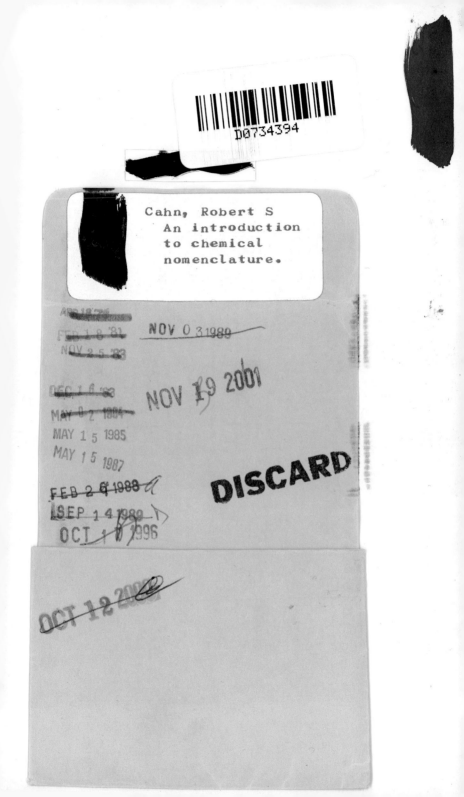

An Introduction to
Chemical Nomenclature

Dedicated to Professor P. E. Verkade, a member of the IUPAC Commission on the Nomenclature of Organic Chemistry for 41 years, for the last 37 years of which as Chairman, who by tact and persistence shepherded the Commission through its national divergencies to ultimate agreements.

An Introduction to
Chemical Nomenclature

R. S. Cahn, M.A., Dr. Phil. nat., F.R.I.C.

FOURTH EDITION

A HALSTED PRESS BOOK

JOHN WILEY & SONS
New York – Toronto

English fourth edition published in 1974 by
Butterworth & Co (Publishers) Ltd
88 Kingsway, London WC2B 6AB

Published in the U.S.A. and Canada by
Halsted Press, a Division of John Wiley & Sons, Inc.,
New York

Library of Congress Cataloging in Publication Data

Cahn, Robert Sidney, 1899–
    An introduction to chemical nomenclature.

    "A Halsted Press book."
    Includes bibliographies.
    1.   Chemisty—Nomenclature.   I.   Title.
QD7.C2   1974   540'.1'4   73-12479
ISBN 0–470–12931–X

Printed in England

Introduction to Fourth Edition

In the fourth edition of this book the whole text has been revised and more than half of it completely rewritten so as to include the new legislation for chemical nomenclature. The inorganic section now includes the essential features of the very valuable new IUPAC recommendations for coordination compounds, π-complexes and boron compounds; and the organic section has been extended to help the reader proceed from principles to actual names. But the essential feature of previous editions is retained, namely to explain the principles of how names are formed rather than to reproduce formal rules.

There has been one change in principle. This edition is written with American conventions, since they are used (though not always correctly) in most treatises or papers in the English language and because they are reproduced in computer outputs of English-language chemical literature. Differences from British conventions are, however, noted where appropriate.

In preparing this edition I have been greatly assisted by Dr. Waldo E. Cohn and Professor W. Klyne with regard to biochemical nomenclature rules, and by Dr. K. Loening through whose kind offices I have been able to consult the 3168-page *Naming and Indexing of Chemical Compounds*, 1969 edition, compiled for use in *Chemical Abstracts Service*. I also thank the Union of Pure and Applied Chemistry and Butterworths for permission to quote freely from the IUPAC nomenclature publications. Only I, however, am to blame for any errors.

R. S. Cahn

Contents

1   The Production, Use and Misuse of Nomenclature

Nomenclature is the tool by which chemists describe their compounds to one another. Like other tools it can be made in different forms and used in different ways, and it can be misused.

Chemists need to describe compounds for various purposes. According to the occasion, a scribbled structural formula, 'That substance', or 'compound (2)' may be the most suitable designation; but names will continue to be needed for lists and legislation, as well as for printed abstracts, indexes and lexicons; and, of course, names are essential, too, for many research papers or reports, for textbooks and for most chemical conversation whether written or spoken.

The rules of chemical nomenclature are as diverse and specialized as the compounds they describe. Over the great complexities and illogicalities of current nomenclature hangs the shadow of the computer; the marshalling of 3–4 million structures and their attendant properties is increasingly admitted to be a computer responsibility, and it is reasonable to expect that the computer language being developed for this task may be adapted for speech and that the resultant words may be superior in logic and accuracy to the often arbitrary and diverse current usage. Such changes have happened before; in spite of their evocativeness and the subsidiary information they contain, names such as muriatic acid and spirits of hartshorn are no longer part of the chemists' vocabulary. However, our new computer language is not yet available, it will not become general parlance at a word of command and, at worst, the present language will for many years still have to be read and understood in existing journals and books. The present nomenclature is thus still worth study; every chemist should know its principles and its correct use.

The nomenclature that is today regarded as 'correct' follows from the consensus of users' opinions. It is written in rule form by the

Commissions of the International Union of Pure and Applied Chemistry and of the International Union of Biochemistry, who try to see nomenclature as a whole, codifying existing practice and occasionally suggesting novelties; they accept the useful practices of specialists within their own fields but reject what they consider to be unnecessary aberrations from general principles.

It would be simpler if there were only one 'correct' name for a substance. In practice, particularly in organic chemistry, this is not so. There are two reasons. First, large compilations such as Beilstein's *Handbuch* and *Chemical Abstracts* often use differing principles, and for them to introduce fundamental changes would bring chaos into their indexes; equally, a single rule is not always practicable, as when two or more large sections of chemists steadfastly maintain different customs. It is, however, noteworthy that all those who formulate rules pay increasing attention to the practices of *Chemical Abstracts*, recognizing the increasing penetration of the Chemical Abstracts Service into every aspect of chemical information. There is one other feature that must be emphasized in a book such as this, devoted mainly to rules, namely that rules are a tool and not a master; they may be overridden or, better, modified when the science or its exposition is thereby improved or made easier to understand.

Because, as just explained, it is not always possible for chemists to agree on the most desirable type of name, there are cases where alternative names are prescribed as equally 'correct' in the international rules. Then one country, Society, journal, or compendium may exercise its own preference. Within reason, each individual chemist has the same choice, though in practice he may be limited by his Society, editor or publisher. In most cases there is one name which is correct for a particular purpose: an author may use one of the alternatives, or even an 'incorrect' name, if it is essential for his theoretical arguments, but not just because of his personal 'preference'; the authorized version will, with a little ingenuity, suffice for almost all purposes.

Now a systematic name for a complex compound is usually itself complex, and some thought will be needed to understand it. It is therefore misuse of nomenclature to scatter long chemical names indiscriminately into a cursive explanation of ideas (necessary though this usually is for the detailed description of an experiment). This misuse of a precision tool is far commoner than it should be: for it is better nomenclature to choose carefully a phrase such as 'the unsaturated alcohol', 'the derived acid', 'the tricyclic ketone', 'the starting material', or simply 'compound (5)' than to bespatter one's prose with names such as 3β-hydroxy-5-oxo-D-nor-5,6-secocholest-

9(11)-en-6-oic acid or 5-(4-diethylamino-1-methylbutyl)dibenz[*aj*]-acridine.

An equally common misuse is false analogy in naming new types or sub-types of compound. Here it is very hard to lay down precise rules. Selection depends on a wide knowledge of previous practice: it is only too easy to mislead—and the overriding criterion for a name is that it shall be unambiguous. The advice of the national expert or editor is here essential.

Difficulties notwithstanding, chemists should, if they wish to be clearly understood, learn to describe accurately the compounds they are handling or talking about—and a definite act of learning is needed. Nomenclature, particularly in its modern developments, is not merely an arbitrary collection of names. It combines past practice with general principles, which it is the object of the following pages to explain. Tampering with it merely makes life harder for the reader and for the searcher in indexes. It is rarely good to call a spade a shovel, with or without a prefix.

2 Inorganic

GENERAL

The greater part of inorganic nomenclature was for many years handled with reasonable ease by means of the endings -ic, -ous, -ium, -ide, -ite and -ate. When these did not suffice, help was sought mainly in prefixes of the type pyro-, hypo-, meta-, ortho-, per-, sub-, and in endings such as -oxylic, -yl and -osyl. There was, however, little consistency in the use of these adjuncts, and the resulting confusion was made worse when later studies of structure disclosed irrationalities in place of some of the supposed analogies. The Stock notation helped in many cases, and Werner's nomenclature was invaluable for coordination compounds.

There have been four recent international attempts to devise a general system for inorganic nomenclature. A comprehensive set of rules was issued[1] by the Commission on the Nomenclature of Inorganic Chemistry of the International Union of Chemistry* in 1940, but because of the war it received no outside comment before publication. A post-war revision was published in 1953 as 'Tentative Rules'[2]; independent comment and further consideration led to 'Definitive' rules[3] resulting from the Paris Conference of 1957. Certain revisions were published by IUPAC in 1965[4]. Finally, in 1971 a new set of Definitive Rules[5] was published by IUPAC that amalgamates, revises and greatly extends previous versions, providing principles, rules and examples over a very wide range. It is on this last set that the present chapter is based. Among the chief features are acceptance of the well-known -ide nomenclature for binary compounds, recommendations for use of either the Stock or the Ewens–Bassett notation, and extension of Werner's system for coordination compounds to a large part of general inorganic

---

* The older title, International Union of Pure and Applied Chemistry (IUPAC), was re-assumed in 1949.

chemistry. Exceptions are still made for very common names such as water or ammonia and for a long list of acids, though the Commission doubtless hopes that these exceptions also will in time be superseded. Much that is familiar remains; and the extensions often lead to familiar or easily recognizable names such as potassium tetrachloroaurate(III) $KAuCl_4$, hydrogen difluorodihydroxoborate $H[B(OH)_2F_2]$ and potassium tetrafluorooxochromate(V) $K[CrOF_4]$; the extensions would, systematically, give disodium tetraoxosulfate for Na_2SO_4 though, of course, sodium sulfate is included among the permitted exceptions. The main virtue of the extension and revision is the replacement of personal or national preference by system and the provision of unambiguous bases for naming new compounds, including many organometallic compounds of great complexity. The only unfortunate aspect is that so many alternatives are left available for, particularly, organometallic coordination compounds.

ELEMENTS

Names and symbols for the elements are given in *Table 2.1* (p. 6). The names in parentheses are those to be used with affixes, e.g. cupric (not copperic), ferrate (not ironate). A few specific points may be noted. Tungsten is now accepted, after an earlier attempt to replace it by wolfram. The symbol for argon is Ar (not A), usage in different countries having become confused and the other noble gases having two-letter symbols. Some compounds of sulfur and antimony are named by use of syllables from the Greek (thion) or Latin (stibium); occasionally old French usage persists in English, as in azide from the French usage of azote for nitrogen. Use of wolframate and niccolate in place of tungstate and nickelate has been recommended. Sulfur, not sulphur, should be used; the English use of sulphur is based on a mistaken belief that sulfur had a Greek origin, in which case ph would replace the Greek phi (φ). The American spelling cesium and aluminum may also be noted.

Some collective names now receive international sanction: noble gases; halogens (F, Cl, Br, I, At); chalcogens (O, S, Se, Te, Po); alkali metals (Li to Fr); alkaline-earth metals (Ca to Ra); lanthanoids for elements 57–71 (La to Lu inclusive) (lanthanides before 1965); actinoids, uranoids and curoids analogously. A transition element is defined as an element whose atoms have an incomplete d-subshell or which gives rise to a cation or cations with an incomplete d-subshell. Compounds of chalcogens are named chalcogenides.

Table 2.1 IUPAC NAMES AND SYMBOLS OF THE ELEMENTS

Name	Symbol	Atomic number	Name	Symbol	Atomic number
Actinium	Ac	89	Mercury	Hg	80
Aluminum*	Al	13	Molybdenum	Mo	42
Americium	Am	95	Neodymium	Nd	60
Antimony	Sb	51	Neon	Ne	10
Argon	Ar	18	Neptunium	Np	93
Arsenic	As	33	Nickel	Ni	28
Astatine	At	85	Niobium	Nb	41
Barium	Ba	56	Nitrogen	N	7
Berkelium	Bk	97	Nobelium	No	102
Beryllium	Be	4	Osmium	Os	76
Bismuth	Bi	83	Oxygen	O	8
Boron	B	5	Palladium	Pd	46
Bromine	Br	35	Phosphorus	P	15
Cadmium	Cd	48	Platinum	Pt	78
Calcium	Ca	20	Plutonium	Pu	94
Californium	Cf	98	Polonium	Po	84
Carbon	C	6	Potassium	K	19
Cerium	Ce	58	Praseodymium	Pr	59
Cesium	Cs	55	Promethium	Pm	61
Chlorine	Cl	17	Protactinium	Pa	91
Chromium	Cr	24	Radium	Ra	88
Cobalt	Co	27	Radon	Rn	86
Copper (Cuprum)	Cu	29	Rhenium	Re	75
Curium	Cm	96	Rhodium	Rh	45
Dysprosium	Dy	66	Rubidium	Rb	37
Einsteinium	Es	99	Ruthenium	Ru	44
Erbium	Er	68	Samarium	Sm	62
Europium	Eu	63	Scandium	Sc	21
Fermium	Fm	100	Selenium	Se	34
Fluorine	F	9	Silicon	Si	14
Francium	Fr	87	Silver (Argentum)	Ag	47
Gadolinium	Gd	64	Sodium	Na	1
Gallium	Ga	31	Strontium	Sr	38
Germanium	Ge	32	Sulfur	S	16
Gold (Aurum)	Au	79	Tantalum	Ta	73
Hafnium	Hf	72	Technetium	Tc	43
Helium	He	2	Tellurium	Te	52
Holmium	Ho	67	Terbium	Tb	65
Hydrogen	H	1	Thallium	Tl	81
Indium	In	49	Thorium	Th	90
Iodine	I	53	Thulium	Tm	69
Iridium	Ir	77	Tin (Stannum)	Sn	50
Iron (Ferrum)	Fe	26	Titanium	Ti	22
Krypton	Kr	36	Tungsten (Wolfram)	W	74
Lanthanum	La	57	Uranium	U	92
Lawrencium	Lr	103	Vanadium	V	23
Lead (Plumbum)	Pb	82	Xenon	Xe	54
Lithium	Li	3	Ytterbium	Yb	70
Lutetium	Lu	71	Yttrium	Y	39
Magnesium	Mg	12	Zinc	Zn	30
Manganese	Mn	25	Zirconium	Zr	40
Mendelevium	Md	101			

* Aluminium is nevertheless still current in British publications and is in accord with the -ium ending adopted for all newly discovered elements.

Compounds of halogens are given as halogenides or halides in the latest IUPAC rules, the latter term being that current in the UK and USA. The term metalloid is vetoed; it is stated that elements should be classified as metallic, semi-metallic or non-metallic, but the idea of a 'semi-metal' may find critics.

Protium, deuterium and tritium are retained for the hydrogen isotopes ^1H, ^2H and ^3H, respectively, but other isotopes should be designated by mass numbers, e.g. oxygen-18 or ^{18}O. The prefixes are deuterio- and tritio- (not deutero-).

Indexes to be used with atomic symbols are:

left upper	. .	mass number
left lower	. .	atomic number
right upper	. .	ionic charge
right lower	. .	number of atoms

For example, $^{32}_{16}S_2^{2+}$ is a doubly charged molecule containing two atoms of sulfur, each atom having the atomic number 16 and mass number 32. These indexes need not all be inserted together; for instance, Ca^{2+} is a doubly ionized calcium ion (with natural abundance of isotopes), ^{15}N an uncharged atom of nitrogen-15, $^{40}K^+$ a singly charged ion of potassium-40. Writing the mass number as upper right index has long been the practice of most physicists, and also of many chemists in America; the physicist often needs to signify atomic attributes other than the four just listed; but the positions given above are best suited for chemists who so often need to cite both mass number and ionic charge.

Radioactivity is often indicated by an asterisk, $^*K$; it is rarely necessary to give both the mass number and the asterisk ($^{*40}K$).

Ionic charge must be given as, e.g. superscript $2+$, and not superscript $+2$.

For allotropic forms of elements a very simple numerical system is recommended: monohydrogen, dioxygen, tetraphosphorus, etc. Trioxygen is then recommended by IUPAC for O_3, though this can hardly be held to exclude use of the familiar name ozone since ozonide is listed among the recognized names of radicals.

Ring and chain structures can be designated by prefixes *cyclo*- and *catena*-, e.g. *cyclo*-octasulfur (or octasulfur; for λ-sulfur), *catena*-sulfur (or polysulfur; for μ-sulfur). A prefix *cyclo*- is now specified for italics by IUPAC in inorganic chemistry (not in organic chemistry).

The description of subgroups of the Periodic Table has been settled by the IUPAC 1965 revision as follows:

settled by the IUPAC 1965 revision as follows:

1A	2A	3A	4A	5A	6A	7A
K	Ca	Sc	Ti	V	Cr	Mn
Rb	Sr	Y	Zr	Nb	Mo	Tc
Cs	Ba	La*	Hf	Ta	W	Re
Fr	Ra	Ac†				

1B	2B	3B	4B	5B	6B	7B
Cu	Zn	Ga	Ge	As	Se	Br
Ag	Cd	In	Sn	Sb	Te	I
Au	Hg	Tl	Pb	Bi	Po	At

* Including the lanthanoids.
† Including the actinoids, but thorium, protactinium and uranium may also be placed in groups 4, 5, and 6.

COMPOUNDS

Formulas and names should correspond to the stoichiometric proportions, expressed in the simplest form which avoids the use of fractions (though semi and sesqui may be used for addition compounds and solvates). The molecular formula, if different, is used only when dealing with discrete molecules whose degree of association is considered independent of temperature. When there is temperature-dependence, the simplest formula is again to be used unless the molecular complexity requires particular emphasis in the context. Thus we have KCl potassium chloride, PCl_3 phosphorus trichloride, S_2Cl_2 disulfur dichloride and $H_4P_2O_6$ hypophosphoric acid (see p. 17); NO_2 nitrogen dioxide represents the equilibrium mixture of NO_2 and N_2O_4 for normal use, but N_2O_4 dinitrogen tetroxide is used where this doubling of the formula is significant.

In *formulas* the electropositive constituent is generally placed first, e.g. PCl_3. But there are exceptions: when clarity demands it, as in NCS^- or CNS^-; when there is a central atom, that should be placed first with the remainder in alphabetical order, as in $PBrCl_2$ and PCl_3O; in acids, H is placed first, as in HCl; and see also the section on radicals (p. 20).

Names of compounds are given in two (or more) parts, the (most) electropositive constituent (cation) first and the (most) electronegative (anion) last. Exceptions are made for neutral coordination compounds, addenda such as solvate molecules (p. 33), and some hydrides (for ozone see p. 7). However, no fundamental distinction is to be made between ionized and non-ionized molecules in general.

Proportions of the various parts are expressed by Greek numerical prefixes (see *Table 3.1*, p. 42; also p. 35); but there are extremely important qualifications that mono (for unity) is usually omitted and that other numerical prefixes may also be omitted if no ambiguity results. Multiplicative numerical prefixes (bis, tris, tetrakis, etc.) are used when followed directly by another numerical prefix and may be used whenever ambiguity might otherwise be caused; and prefixes may be delimited by parentheses to aid clarity further (examples are on pp. 24, 25). The terminal 'a' of tetra, penta, etc., has normally been elided in English before another vowel in inorganic chemistry, but this is now expressly forbidden in IUPAC inorganic nomenclature.

BINARY COMPOUNDS

Compounds between two elements are called binary compounds, independently of the number of atoms of each element in a molecule; e.g. they include N_2O, NO, NO_2, N_2O_4, etc.

In formulas and names of compounds between two non-metals that constituent is placed first which occurs earlier in the sequence:

Rn, Xe, Kr, B, Si, C, Sb, As, P, N, H, Te, Se, S, At, I, Br, Cl, O, F

This order is arbitrary in places: it is not based solely on an order of electronegativity. The results are mostly familiar: NH_3 (not H_3N), CCl_4, NCl_3, NO, etc. But Cl_2O (chlorine monoxide) contrasts with O_2F (dioxygen fluoride).

When neither atom of a binary compound occurs in the sequence Rn . . . F above, the atoms are cited in the inverse order of the element sequence shown in *Table 2.2* (p. 10). This applies both to formulas and names, e.g. Na_2Pb, sodium plumbide.

Metals precede non-metals, in formulas and names.

When forming a complete name one leaves the name of the electropositive constituent unmodified, except when it is necessary to indicate the valence or oxidation state (see below); and the same applies to the name of that one of a pair of non-metals which is cited first in the list Rn . . . F. The name of the electronegative constituent, or of the non-metal named second, is modified to end in -ide. Normally this modification is carried out by stripping the name of the element back to the penultimate consonant and then adding 'ide', as in carbon, carb, carbide; or chlorine, chlor, chloride; but the following exceptions should be noted:

bismuth, bismuthide;	hydrogen, hydride;
mercury, mercuride;	nitrogen, nitride;
oxygen, oxide;	phosphorus, phosphide;
	zinc, zincide

Table 2.2 ELEMENT SEQUENCE

When Latin names are given in *Table 2.1*, these are used, in the normal way.

Coupling this rule with those for numerals (pp. 8–9), and not forgetting the provisions for omission of numerical prefixes, we then have a host of familiar names, of which the following list, based on one in the IUPAC rules, is representative: sodium chloride, calcium sulfide, lithium nitride, arsenic selenide, calcium phosphides, nickel arsenide, aluminum borides, iron carbides, boron hydrides, phosphorus hydrides, hydrogen chloride, hydrogen sulfide, silicon carbide, carbon disulfide, sulfur hexafluoride, chlorine dioxide, oxygen difluoride, sulfur dioxide, sulfur trioxide, carbon monoxide, carbon dioxide, dinitrogen oxide (N_2O), nitrogen oxide (NO), dinitrogen pentaoxide (N_2O_5).

This list will be seen to cover both ionized and un-ionized compounds. (Note the hyphen used to distinguish un-ionized from union-ized.) The compounds HHal are called hydrogen chloride, hydrogen bromide, hydrogen iodide and hydrogen fluoride: the names hydrochloric acid, etc., refer to aqueous solutions, and percentages such as 20% hydrochloric acid denote the wt./vol. of hydrogen halide in the solution (not the dilution of the concentrated acid by water). Also, hydrogen azide is recommended by IUPAC rather than hydrazoic acid.

Now, oxides could be, but are not, named by the coordination principle (see p. 14): e.g. SO_3 is sulfur trioxide (not trioxosulfur). The same applies to peroxides, i.e. compounds whose electronegative constituent is the group O_2^{2-} (when the electronegative constituents are atoms 2O or ions $2O^-$ the compounds are dioxides); for example, Na_2O_2 sodium peroxide (but PbO_2 lead dioxide).

Binary hydrides are, with certain exceptions, named by adding -ane to the root name of the non-hydrogen element, e.g. borane (BH_3), stannane (SnH_4). If the number of atoms of the non-hydrogen element in the molecule exceeds one, this is indicated by a Greek numerical prefix, as in diborane (B_2H_4), trisilane (Si_3H_8) and pentasulfane (H_2S_5). There is, however, a short list of exceptions hallowed by long usage, namely: water, ammonia, hydrazine, phosphine, arsine, stibine and bismuthine. Also hydrides of Group VII, being acidic, are named as hydrogen halides (cf. above). Occasionally hydride names ending in -ane may become confused with the organic termination -ane for a six-membered ring; for this reason, IUPAC recommends diselane H_2Se_2 and ditellane H_2Te_2 and their homologs.

Many trivial names have been abandoned in modern nomenclature, often owing to mere omission from IUPAC rules; notable cases are the oxides of nitrogen, but, except for custom and sentiment among the older chemists, they are little loss.

For most compounds, in most circumstances, it is unnecessary to specify valence or oxidation state. It can always be dispensed with when no ambiguity results, as can the numerical prefixes. So one still has very many simple names such as sodium chloride, calcium oxide, deuterium oxide and barium chloride.

However, there are far more cases where such simplicity leaves ambiguous names and at this stage the concept of oxidation number must be stressed. It is expressed in the 1970 IUPAC rules as follows:

The oxidation number of an element in any chemical entity is the charge which would be present on an atom of the element if the electrons in each bond to that atom were assigned to the more electronegative atom, thus:

Oxidation numbers

MnO_4^- = one Mn^{7+} and four O^{2-} ions	$Mn = VII$	$O = -II$
ClO^- = one Cl^+ and one O^{2-} ion	$Cl = I$	$O = -II$
CH_4 = one C^{4-} and four H^+ ions	$C = -IV$	$H = I$
CCl_4 = one C^{4+} and four Cl^- ions	$C = IV$	$Cl = -I$
NH_4^+ = one N^{3-} and four H^+ ions	$N = -III$	$H = I$
NF_4^+ = one N^{5+} and four F^- ions	$N = V$	$F = -I$
AlH_4^- = one Al^{3+} and four H^- ions	$Al = III$	$H = -I$
$[PtCl_2(NH_3)_2]$ = one Pt^{2+} and two Cl^- ions and two uncharged NH_3 molecules	$Pt = II$	$Cl = -I$
$[Ni(CO)_4]$ = one uncharged Ni atom and four uncharged CO molecules	$Ni = 0$	

By convention hydrogen is considered positive in combination with non-metals. The conventions concerning the oxidation numbers of organic radicals and the nitrosyl group are considered below.

In the elementary state the atoms have oxidation state zero and a bond between atoms of the same element makes no contribution to the oxidation number, thus:

Oxidation numbers

P_4 = four uncharged P atoms	$P = 0$	
P_2H_4 = two P^{2-} and four H^+ ions	$P = -II$	$H = I$
C_2H_2 = two C^- and two H^+ ions	$C = -I$	$H = I$
O_2F_2 = two O^+ and two F^- ions	$O = I$	$F = -I$
$Mn_2(CO)_{10}$ = two uncharged Mn atoms and ten uncharged CO molecules	$Mn = 0$	

When given elements can combine in different proportions the names describing the resulting different compounds may be formed in four ways:

(i) The first way is to use numerical prefixes that show the stoichiometric composition, as in iron trichloride, copper dichloride, triiron tetraoxide and dinitrogen pentasulfide.

(ii) In the second method the oxidation number, in Roman numerals in parentheses, is placed immediately after the name of the element concerned (but the zero is denoted by the Arabic 0). This is the Stock system which is widely favored. Examples are:

$$MnO_2 \quad \text{manganese(\textsc{iv}) oxide}$$
$$BaO \quad \text{barium(\textsc{ii}) oxide}$$

The rules list both English and Latin names, but favor the latter where both are given in *Table 2.1*; for instance:

$$FeCl_3 \quad \text{iron(\textsc{iii}) chloride or ferrum(\textsc{iii}) chloride}$$
$$CuCl_2 \quad \text{copper(\textsc{ii}) chloride or cuprum(\textsc{ii}) chloride}$$

(iii) Alternatively the charge on an ion may be indicated by an Arabic numeral, followed by the sign of the charge, both in parenthesis attached at the end of the name of the ion (the Ewens–Bassett system). Examples are:

$$FeCl_3 \quad \text{iron(3+) chloride}$$
$$CuCl_2 \quad \text{copper(2+) chloride}$$

(iv) There is of course also the much older system of denoting a higher valence state by the ending -ic, and a lower one by the ending -ous. This system is notoriously incomplete when an element can exist in more than two valence states, is not applicable to coordination compounds and can be a troublesome tax on memory for the less common elements. The IUPAC Commission lists the system as 'in use but discouraged', although there is also a statement that it 'may be retained for elements exhibiting not more than two valencies'.

Here we may note that it is only in connection with the -ous/-ic and -ite/-ate terminations that the word 'valency' is used in the 1970 IUPAC rules; elsewhere their rules are based on oxidation numbers and ionic charges.

PSEUDO-BINARY COMPOUNDS

This name has been applied to compounds containing some common polyatomic anions that customarily have names ending in -ide; these anions are:

HO^-	hydroxide ion	N_3^-	azide ion
O_2^{2-}	peroxide ion	NH^{2-}	imide ion
O_2^-	hyperoxide ion	NH_2^-	amide ion
O_3^-	ozonide ion	$NHOH^-$	hydroxylamide ion
S_2^{2-}	disulfide ion	$N_2H_3^-$	hydrazide ion
I_3^-	triiodide ion	CN^-	cyanide ion
HF_2^-	hydrogen difluoride ion	C_2^{2-}	acetylide ion

Treating these as the other -ide names we obtain NaOH sodium hydroxide, KCN potassium cyanide and the like. Sodium amide ($NaNH_2$) is often abbreviated to sodamide, but there is no need for this practice.

The list above includes S_2^{2-} as the disulfide ion and there are clearly homologous ions; but in this series it has become the practice to name the chain compounds $HS–[S]_x–SH$ as sulfanes ($x = 1$, trisulfane; $x = 2$, tetrasulfane; etc.); this is particularly valuable for organic derivatives and for hydroxy, amino, etc., derivatives, but it is then probably wisest to name the substituents only as prefixes rather than to encounter the difficulties in deciding a choice of suffix.

THE EXTENDED COORDINATION PRINCIPLE

In its oldest sense, the sense in which it is usually understood, a coordination compound is one containing an atom (A) directly attached to other atoms (B) or groups (C), or both, the number of these being such that the oxidation number or stoichiometric valence of the atom (A) is exceeded.

The 1957 IUPAC rules extended this principle by abolishing the restriction that the oxidation number must be exceeded. The effect is to bring the major part of nomenclature potentially within the scope of the nomenclature customary for coordination compounds. For instance, in the group SO_4, the sulfur atom is atom (A) and the oxygen atoms are (B). Even for diatomic groups such as ClO^- a 'characteristic' atom (here Cl) can generally be recognized and the remainder of the nomenclature welded to that.

For discussion of this nomenclature some definitions should be noted (some will not be needed until later in this chapter, but it is convenient to group them together):

Central or nuclear atom: the atom (A) above.

Coordinating atom: each atom *directly* attached to the central atom.

Coordination number or ligancy: the number of atoms directly attached to a central atom.

Ligand: each atom (B) or group (C) attached to the central atom.

Multidentate ligand: a group containing more than one *potential* coordinating atom. Hence, unidentate, bidentate, etc.

Chelate ligand: a ligand actually attached to one central atom through two or more coordinating atoms.

Bridging group: a ligand attached to more than one central atom.

Complex: the whole assembly of one or more central atoms with their attached ligands.

Polynuclear complex: a complex containing more than one central (nuclear) atom. Hence, mononuclear, dinuclear, etc.

To name the complex, the names of the ligands are attached directly in front of the name of the central atom; the oxidation number of the central atom is then stated last, or the proportions of the constituents are indicated by stoichiometric numerical prefixes; when the oxidation number is exceeded, both the numerical prefixes and the oxidation number are often needed.

The complex may be cationic, neutral, or anionic. Names of cationic and neutral complexes are not modified (except for a few mentioned below), i.e. they have no characteristic ending. Names of anionic complexes are given the ending -ate, often with abbreviation.

It is very necessary to realize that according to this system the ending -ate denotes *merely* an anionic complex. When an element has variable oxidation state this use of -ate does *not* denote a higher oxidation state; it is used with all oxidation states, the distinction between them being made by citing the oxidation number. Particular note must be taken of this because the -ate, -ite, -ide distinction is still permitted in a limited and defined, but large, number of cases, and the two nomenclatures are likely to be in use side by side for some time.

The ligands are to be cited in alphabetical order*. In this alphabetization, multiplying prefixes are neglected; e.g. di-iodo is alphabetized under i and thus after fluoro, but both are placed after trichloro which is alphabetized under c (for further details see p. 55).

The names of anionic ligands are changed to end in 'o' (see p. 24); those of neutral and cationic ligands are unchanged, except that H_2O is denoted aqua† and NH_3 ammine.

---

* The 1957 rules gave complex instructions about the order of citation of ligands. All that is now swept away in favor of the alphabetical order.

† Aqua is a change from aquo which was customary before 1965.

ACIDS AND NORMAL SALTS CONTAINING MORE THAN TWO ELEMENTS

Classical coordination compounds, that is, those where the oxidation number of the central atom is exceeded, are deferred to p. 24.

(a) The practices mentioned earlier hold, about citing electropositive before electronegative constituents, using Stock numbers or -ic, -ous endings, omitting numerical prefixes, or the Ewens–Bassett system, or Stock numbers and charges when no confusion arises.

For acids containing more than two elements the IUPAC recommendations of 1970 are somewhat more permissive than those of 1959. The latter held out its hope that the nomenclature typified by sulfuric acid would be replaced by hydrogen sulfate and the like. The 1970 rules freely permit both types, while, however, expressing preference for the hydrogen -ate type for the less common acids.

Both sets of rules list common acids for which the 'acid' nomenclature is likely to be retained. *Table 2.3* is the more extensive 1970 list.

Anions from the acids in *Table 2.3* are formed by changing -ous acid to -ite, and -ic acid to -ate, and this gives the familiar names for a further host of common compounds. For instance, $NaNO_2$ sodium nitrite and $NaNO_3$ sodium nitrate follow at once. Na_2SO_4 sodium sulfate and Na_2CO_3 sodium carbonate follow similarly when permission to omit unnecessary prefixes (e.g. from disodium sulfate) is remembered.

A few general principles for the use of prefixes embodied in *Table 2.3* may be pointed out. A prefix hypo- denotes a lower oxidation state and is used with the -ous and the -ic termination. Per- denotes a higher oxidation state, is used only with the -ic acids (and their salts), and must not be confused with peroxo- (see below). Ortho- and meta- distinguish acids of differing water content.

Thio acids are acids derived by replacing oxygen by sulfur, and are named by adding thio- before the trivial name of the acid; and seleno and telluro acids are treated similarly. Examples are: $H_2S_2O_2$ thiosulfurous acid; $H_2S_2O_3$ thiosulfuric acid; KSCN potassium thiocyanate; H_3PO_3S monothiophosphoric acid; $H_3PO_2S_2$ dithiophosphoric acid; H_2CS_3 trithiocarbonic acid. However, $H_2S_2O_6$ dithionic acid and $H_2S_2O_4$ dithionous acid (not hydrosulfurous or hyposulfurous) in *Table 2.3* are exceptional to this treatment, and higher homologs are named analogously.

Peroxo acids, in which –O– is replaced by –O–O–, are similarly distinguished by the prefix peroxo- (peroxy- or simply per- have frequently been used in the past), as, for example, in HNO_4 peroxonitric acid and K_2SO_5 potassium peroxosulfate, HO–O–NO peroxonitrous acid and $H_4P_2O_8$ peroxodiphosphoric acid.

(b) To explain how acids and salts are named by the coordination method requires, first, the principles given in the preceding section and then further precision about endings.

It will be remembered that the electropositive portion is named before the electronegative, as usual, and that the electronegative, anionic portion has an ending -ate. This ending is normally added to the penultimate consonant of the name of the central atom (the

Table 2.3 NAMES FOR ACIDS CONTAINING MORE THAN TWO ELEMENTS*

H_3BO_3	orthoboric acid† or boric acid	H_2SO_4	sulfuric acid
$(HBO_2)_n$	metaboric acid	$H_2S_2O_7$	disulfuric acid
H_2CO_3	carbonic acid	H_2SO_5	peroxomonosulfuric acid
HOCN	cyanic acid		
HNCO	isocyanic acid	$H_2S_2O_8$	peroxodisulfuric acid
HONC	fulminic acid	$H_2S_2O_3$	thiosulfuric acid
H_4SiO_4	orthosilicic acid†	$H_2S_2O_6$	dithionic acid
$(H_2SiO_3)_n$	metasilicic acid	H_2SO_3	sulfurous acid
HNO_3	nitric acid	$H_2S_2O_5$	disulfurous acid
HNO_4	peroxonitric acid	$H_2S_2O_2$	thiosulfurous acid
HNO_2	nitrous acid	$H_2S_2O_4$	dithionous acid
HOONO	peroxonitrous acid	H_2SO_2	sulfoxylic acid
H_2NO_2	nitroxylic acid	$H_2S_xO_6$	polythionic acids
$H_2N_2O_2$	hyponitrous acid	(x = 3,4...)	
H_3PO_4	orthophosphoric† or phosphoric acid	H_2SeO_4	selenic acid
		H_2SeO_3	selenious acid
$H_4P_2O_7$	diphosphoric or pyrophosphoric acid	H_6TeO_6	orthotelluric acid†
		H_2CrO_4	chromic acid
		$H_2Cr_2O_7$	dichromic acid
$(HPO_3)_n$	metaphosphoric acid	$HClO_4$	perchloric acid
		$HClO_3$	chloric acid
H_3PO_5	peroxomonophosphoric acid	$HClO_2$	chlorous acid
		HClO	hypochlorous acid
$H_4P_2O_8$	peroxodiphosphoric acid	$HBrO_4$	perbromic acid
		$HBrO_3$	bromic acid
$(HO)_2OP-PO(OH)_2$	hypophosphoric acid or diphosphoric(IV) acid	$HBrO_2$	bromous acid
		HBrO	hypobromous acid
$(HO)_2P-O-PO(OH)_2$	diphosphoric(III,V) acid	H_5IO_6	orthoperiodic acid†
		HIO_4	periodic acid
		HIO_3	iodic acid
H_2PHO_3	phosphonic acid	HIO	hypoiodous acid
$H_2P_2H_2O_5$	diphosphonic acid	$HMnO_4$	permanganic acid
HPH_2O_2	phosphinic acid	H_2MnO_4	manganic acid
H_3AsO_4	arsenic acid	$HTcO_4$	pertechnetic acid
H_3AsO_3	arsenious acid	H_2TcO_4	technetic acid
$HSb(OH)_6$	hexahydroxo-antimonic acid	$HReO_4$	perrhenic acid
		H_2ReO_4	rhenic acid

* Certain other thio and peroxo acids, as well as seleno and telluro acids, can be named similarly.
† Ortho- need be used only when distinction from other acids is essential.

Latin name being used if one is given in *Table 2.1*), as with the -ide ending; but there are the following exceptions: antimony, antimonate; bismuth, bismuthate; carbon, carbonate; cobalt, cobaltate; nickel, nickelate*; nitrogen, nitrate; phosphorus, phosphate; tungsten, tungstate; zinc, zincate.

Names for ligands (which in the simple acids now considered are all anionic) are formed by changing the endings -ide, -ite or -ate to -ido, -ito or -ato, respectively, for direct union to the name of the central atom; the following do not conform but are retained in deference to common older usage:

F^-	fluoro	S^{2-}	thio
Cl^-	chloro	(but: S_2^{2-}	disulfido)
Br^-	bromo	HS^-	mercapto
I^-	iodo	CN^-	cyano
O^{2-}	oxo	CH_3O^-	methoxo‡ or methanolato
H^-	hydrido or hydro†	CH_3S^-	methylthio or
OH^-	hydroxo		methanethiolato
O_2^{2-}	peroxo‡		
HO_2^-	hydrogenperoxo		

In this way $H_2[SO_4]$ becomes dihydrogen tetraoxosulfate (vi); or, since the 'acid' terminology is also permitted, this might become tetraoxosulfuric acid. If we were to make the further provision that oxo- prefixes might be omitted when no confusion is caused, we should reach the familiar sulfuric acid; nevertheless, such argument can lead to confusion in some cases, so it is better to regard the name sulfuric acid simply as one of the permitted exceptions.

The majority of simple oxo acids are included in *Table 2.3*. Note the names orthoperiodic acid for H_5IO_6 and periodic acid for HIO_4, which resolve previous differences of opinion.

Some examples of real usefulness of the method are:

$HReO_4$	tetraoxorhenic(vii) acid
H_3ReO_5	pentaoxorhenic(vii) acid
H_2ReO_4	tetraoxorhenic(vi) acid
$HReO_3$	trioxorhenic(v) acid
$H_4Re_2O_7$	heptaoxodirhenic(v) acid
H_3GaO_3	gallic(iii) acid
H_4XeO_6	hexaoxoxenonic(viii) acid

* IUPAC rules recommend niccolate.

† Both hydrido and hydro are used for coordinated hydrogen but the latter term usually is restricted to boron compounds.

‡ In conformity with the practice of organic nomenclature, the forms peroxy and methoxy are also used but are not recommended.

In simpler cases, it seems reasonable to omit the oxo- prefixes, as in manganic(VI) acid for H_2MnO_4, and manganic(V) acid for H_3MnO_4.

So far as simple compounds are concerned, it is perhaps with halo acids that the Stock nomenclature comes best into its own. Potassium hexachloroplatinate(IV) for K_2PtCl_6 sidesteps the older platini-chloride. Chlorosulfuric(VI) acid for $ClSO_3H$, where Cl could be considered to replace OH of sulfuric acid, avoids confusion with organic chemistry where replacement is normally 'substitution', i.e. replacement of hydrogen. On this system the hydrogen of acids is named as the cation, and it is this principle that leads to hydrogen -ate names. Some further simple examples may be here adduced as illustrations.

$K\ AuCl_4$	potassium tetrachloroaurate
$Na\ [PHO_2F]$	sodium fluorohydridodioxophosphate
$H\ PF_6$	hydrogen hexafluorophosphate
$H_4\ XeO_6$	hydrogen hexaoxoxenonate(VIII) (cf. p. 18)
$H\ [B(OH)_2F_2]$	hydrogen difluorodihydroxoborate
$Na_4\ [Fe(CN)_6]$	sodium hexacyanoferrate(II)
$K\ [AgF_4]$	potassium tetrafluoroargentate(III)
$Ba\ [BrF_4]_2$	barium tetrafluorobromate(III)
$K\ [Au(OH)_4]$	potassium tetrahydroxoaurate(III)
$Na\ [AlCl_4]$	sodium tetrachloroaluminate
$Li\ [AlH_4]$	lithium tetrahydridoaluminate

The last example in that list may come as a shock to some organic chemists, for it replaces the older and less precise lithium aluminum hydride.

Some further points in the above examples may also repay comment:

The most recent IUPAC rules recommend that in formulas the square brackets indicating the extent of a complex ion be separated by a space from any other chemical symbol or another set of square brackets. It remains to be seen whether chemists in general favor this or regard it as painting the lily—after all, the square brackets themselves define the extent of the complex ion.

The prefix hydrido- denoting a hydride anionic ligand (cf. p. 18) is especially noteworthy; it follows the general rule of changing an ending -ide to -ido and for inorganic chemistry is considered pre-ferable to the organic usage of hydro. A similar argument could have been applied to the other exceptional anionic ligand prefixes listed on p. 18, e.g. chlorido- in place of chloro-, but here the need to alter previous usage was not so necessary as with hydrogen which so often occurs as a proton H^+.

Lastly, it may be well to anticipate here a little from later pages by mentioning that hydrocarbon radicals that are present as ligands do not receive the terminal -o; so we have, for example, Na [B(C$_6$H$_5$)$_4$] sodium tetraphenylborate, which will be of interest also to the organic chemist.

IONS AND RADICALS

Anions are negatively, and cations positively, charged atoms or groups of atoms. Radicals are uncharged groups of atoms that occur (not necessarily always filling the same role) throughout a number of compounds and do not normally exist in the free state; when they exist independently and uncharged they are 'free radicals'; when charged they become ions. Names of ions are, in general, those of the electropositive or electronegative 'constituents' described in the preceding sections; names of cations thus have no characteristic ending; those of anions end in -ide, -ite or -ate. Thus Na$^+$ is the sodium ion or, more precisely, the sodium cation. Fe^{2+} is the ferrous ion, or iron(II) ion, or ferrum(II) ion (or cation may replace ion). I$^-$ is the iodide ion or iodide anion, so it obviously is not wise to abbreviate the name of the less common I$^+$ iodine cation.

There are a few special points to be noted, particularly where the IUPAC rules choose between previous alternative usages.

NO$^+$ is to be called the nitrosyl cation, NO$_2^+$ the nitryl cation (not nitroxyl, to avoid confusion with the radical from nitroxylic acid).

Polyatomic cations formed by adding more protons to monoatomic anions than are required to give a neutral unit have the ending -onium: ammonium, phosphonium, arsonium, oxonium (H$_3$O$^+$), sulfonium, selenonium, telluronium, iodonium. Substituted derivatives may be formed from them, e.g. hydroxylammonium, tetramethylstibonium, dimethyloxonium (CH$_3$)$_2$OH$^+$.

For nitrogen bases other than ammonia and its substitution products the cations are named by changing the final -e to -ium: anilinium, imidazolium, glycinium, etc. N$_2$H$_5^+$ is hydrazinium(1+); N$_2$H$_6^{2+}$ is hydrazinium(2+). Uronium and thiouronium (from urea and thiourea) are exceptions. Names such as dioxanium and acetonium are formed analogously.

Finally, though H$^+$ is the proton, and H$_3$O$^+$ (a monohydrated proton) is oxonium, the term hydrogen ion can be used when the degree of hydration is of no importance in the particular circumstances.

Radicals having special names are listed in *Table 2.4*. Thio-, seleno-, etc., prefixes may be used with these, as with acids. The '-yl' principle is noted in the rules as not extensible to other metal–oxygen radicals. In some cases use of Stock numbers or the Ewens–Bassett system extends the range of utility of these special radical names, e.g. UO_2^{2+} uranyl(VI) or uranyl(2+), UO_2^+ uranyl(V) or uranyl(1+).

Table 2.4 SPECIAL INORGANIC RADICAL NAMES

HO	hydroxyl	SeO	seleninyl
CO	carbonyl	SeO_2	selenonyl
NO	nitrosyl	CrO_2	chromyl
NO_2	nitryl	UO_2	uranyl
PO	phosphoryl	NpO_2	neptunyl
SO	sulfinyl (thionyl)	PuO_2	plutonyl*
SO_2	sulfonyl (sulfuryl)	ClO	chlorosyl†
S_2O_5	disulfuryl	ClO_2	chloryl†
		ClO_3	perchloryl†

* Similarly for other actinides.
† Similarly for other halogens.

These radical names can be used to construct compound names such as:

$COCl_2$	carbonyl chloride
PON	phosphoryl nitride
$PSCl_3$	thiophosphoryl chloride
CrO_2Cl_2	chromyl chloride
IO_2F	iodyl fluoride
SO_2NH	sulfonyl imide or sulfurylimide

It will be seen that these -yl radical names are always handled as if they were the electropositive part of a name, but no polarity considerations should in fact apply: NOCl is nitrosyl chloride, and $NOClO_4$ is nitrosyl perchlorate independently of views on polarity.

Analogous radicals in which a chalcogen replaces oxygen are named analogously, e.g. thiophosphoryl PS.

The radical names, then, are convenient for certain groups of compound, but must be restricted to compounds consisting of discrete molecules, and other, superficially similar compounds are often better named as mixed oxides or oxide salts (see p. 23).

The IUPAC rules use these radicals also for amides, giving $SO_2(NH_2)_2$ sulfonyl diamide and $PO(NH_2)_3$ phosphoryl triamide. As alternatives they give sulfuric diamide and phosphoric triamide. For the partial amides they put forward -amidic acid names, and as an alternative the use of coordination nomenclature; for example, NH_2SO_3H sulfamidic acid or amidosulfuric acid. Abbreviated names—in the above cases the common sulfamide, phosphamide and sulfamic acid—are 'not recommended'. However, the -yl amide names differ noticeably from present organic practice, where amides are a much larger class than in inorganic chemistry; it seems to the present writer better to use either the trivial names or the coordination names at present: for example, $SO_2(NH_2)_2$ sulfamide; or, for new compounds, as in diaminodioxosulfur.

SALTS AND SALT-LIKE COMPOUNDS

(a) 'Acid' salts, i.e. salts containing acid hydrogen

'Names are formed by adding the word "hydrogen", with numerical prefix where necessary, to denote the replaceable hydrogen in the salt. Hydrogen shall be followed without space by the name of the anion. Exceptionally, inorganic anions may contain hydrogen which is not replaceable. It is still designated by hydrogen, if it is considered to have the oxidation number $+1$, but the salts cannot of course be called acid salts.

Examples:

1.	$NaHCO_3$	sodium hydrogencarbonate
2.	LiH_2PO_4	lithium dihydrogenphosphate
3.	KHS	potassium hydrogensulfide
4.	$NaHPHO_3$	sodium hydrogenphosphonate'

That is a direct transcription from the 1970 IUPAC rules. The citation of acidic hydrogen is classical nomenclature, but the junction of two words (hydrogen and an anion) that can have independent existence, without any form of punctuation, is uncommon in British or American nomenclature. Its justification is that the primary ionization is to, e.g. Na^+ and $[HCO_3]^-$ and the custom may perhaps become popular.

It is to be noted that names such as bicarbonate and bisulfate are not permitted; nothing similar is anyway possible for polybasic acids; and the word hydrogen should always be cited when this atom is present. Derived ions should be similarly named, e.g. hydrogencarbonate ion or hydrogen carbonate ion.

Example 4 in the list above, sodium hydrogenphosphonate, emphasizes that the name phosphonic acid is given to H_2PHO_3 (see *Table 2.3*, p. 16) although this acid was still given the older name phosphorous acid in the 1957 IUPAC rules. Phosphonic acid is in line with organic nomenclature such as phenylphosphonic acid for $C_6H_5PO_3H_2$, but it should be noted that at the time of writing no published IUPAC rules are available for detailed nomenclature of the extensive phosphorus chemistry.

(b) Double salts, etc.

A somewhat complex set of IUPAC rules published in 1957 has since been much simplified by extensive use of the alphabetical order.

In formulating and in naming double salts, etc., all the cations are now to be cited first, in alphabetical order, except that H^+ is named last. Then all the anions are listed, these also in their alphabetical order. Water considered to be coordinately bound to a specific ion is cited as aqua- (note the change from aquo- that was customary before 1965). Numerical prefixes can be omitted if the oxidation states are constant or stated. Examples are:

$TlNa(NO_3)_2$ sodium thallium(I) nitrate or
                   sodium thallium dinitrate
$NaZn(UO_2)_3(C_2H_3O_2)_9 \cdot 6H_2O$ sodium triuranyl zinc
                          acetate hexahydrate
$Na_6ClF(SO_4)_2$ (hexa)sodium chloride fluoride bis(sulfate)
$[Cr(H_2O)_6]Cl_3$ hexaaquachromium(III) chloride or
                  hexaaquachromium trichloride

In the penultimate example, either one of the numeral prefixes may be omitted. Note, too, the use of bis(sulfate) to avoid confusion with disulfate ($S_2O_7^{2-}$).

(c) 'Basic' salts

These are to be named as double salts with O^{2-} or OH^- anions; names such as hydroxychloride join the constituents together incorrectly and should not be used (in English). The IUPAC class names are oxide salts and hydroxide salts, e.g.

         $Mg(OH)Cl$ magnesium chloride hydroxide
         $BiOCl$ bismuth chloride oxide
         $Cu_2(OH)_3Cl$ dicopper chloride trihydroxide

(d) Double oxides and hydroxides

These are named on similar principles, the metals being cited in alphabetical order. Examples are:

$FeTiO_3$ iron(II) titanium trioxide
$Cu(CrO_2)_2$ chromium(III) copper(II) oxide
$AlLiMn^{IV}_2O_4(OH)_4$ aluminum lithium dimanganese(IV)
                        tetrahydroxide tetraoxide

(Copper chromite is considered incorrect for the second example in view of results of structural studies on the solid.)

CLASSICAL COORDINATION COMPOUNDS

These are the compounds where the oxidation number is exceeded. Most of the principles governing their nomenclature have been explained above. The main features may be recapitulated as follows. The complex may be cationic, neutral or anionic. Cationic or neutral complexes have no special ending (except -ium for quaternary ions); names of anionic complexes end in -ate. Ligands are cited in the order (i) anionic, alphabetically, (ii) neutral and cationic ligands alphabetically. Stock numbers or Ewens–Bassett charge numbers are cited when necessary or helpful, but stoichiometric prefixes may in some cases suffice alone.

Cationic and neutral ligands have unaltered names; anionic ligands are given the ending -o.

These and the following additional principles suffice for a very large number of coordination compounds:

(a) Hydrocarbon radicals have the usual organic names ending in -yl (not -ylo), but they must however be considered as anions when undesirable oxidation numbers are otherwise obtained.

(b) CH_3O^-, etc., are called methoxo, etc.

(c) CH_3S^- is named methanethiolato or methylthio.

(d) When an organic compound that is not normally named as an acid forms a ligand *by loss of a proton* it is treated as anionic and its name is given the ending -ato. (In other cases it is treated as neutral.) When necessary the charge on such a ligand is to be stated, e.g. $^-OOCCH(O^-)CH(OH)COO^-$ tartrato(3−) and $^-OOCCH(OH)-CH(OH)COO^-$ tartrato(2−).

(e) For computation of the oxidation number, NO, NS, CO and CS are treated as neutral.

(f) Water and ammonia as neutral ligands are designated aqua and ammine, respectively.

(g) When a ligand might be attached at different points, distinction is made by means of letters -*S*, -*N*, -*O*, etc., at the end of the name, unless the customary name makes this clear (e.g. –SCN thiocyanato, –NCS isothiocyanato).

(h) Names of neutral ligands, other than H_2O, NH_3, NO, NS, CO and CS, are placed in parentheses.

(i) Enclosing marks are used in the sequence {[( )]} for composite groups, as necessary, but the order is [{( )}] if the outside square brackets define the complex.

After these expositions the following examples will, it is hoped, be self-explanatory and sufficently illustrate these principles of the alternative systems. Bold type illustrates operation of the alphabetical order (for details see p. 55).

Na[B(NO₃)₄]
sodium tetranitratoborate(1−)
sodium tetranitratoborate(III)

[CoN₃(NH₃)₅] SO₄
penta**ammineaz**idocobalt(2+) sulfate
penta**ammineaz**idocobalt(III) sulfate

[Ru(HSO₃)₂(NH₃)₄]
tetra**ammine**bis(**h**ydrogensulfito)ruthenium
tetra**ammine**bis(**h**ydrogensulfito)ruthenium(II)

K[B(C₆H₅)₄]
potassium tetraphenylborate(1−)
potassium tetraphenylborate(III)

[Fe(C₂C₆H₅)₂(CO)₄]
tetra**c**arbonylbis(**p**henylethynyl)iron
tetra**c**arbonylbis(**p**henylethynyl)iron(II)

[Ni(C₄H₇N₂O₂)₂] .
bis(2,3-butanedione dioximato)nickel
bis(2,3-butanedione dioximato)nickel(II)

[CoCl₂(C₄H₈N₂O₂)₂]
bis(2,3-**b**utanedione dioxime)di**c**hlorocobalt
bis(2,3-**b**utanedione dioxime)di**c**hlorocobalt(II)

[CuCl₂(CH₃NH₂)₂]
di**c**hlorobis(**m**ethylamine)copper
di**c**hlorobis(**m**ethylamine)copper(II)

[Pt(py)₄] [PtCl₄]
tetrakis(pyridine)platinum(2+) tetrachloroplatinate(2−)
tetrakis(pyridine)platinum(II) tetrachloroplatinate(II)

[Cr(H₂O)₆]Cl₃
hexaaquachromium(3+) chloride
hexaaquachromium trichloride

[Co(NH₃)₆]Cl(SO₄)
hexaamminecobalt(3+) chloride sulfate
hexaamminecobalt(III) chloride sulfate

[CoCl₃(NH₃)₂{(CH₃)₂NH}]
diamminetrichloro(**d**imethylamine)cobalt
diamminetrichloro(**d**imethylamine)cobalt(III)

K[Co(CN)(CO)₂(NO)]
potassium di**carbonylcyano**nitrosylcobaltate(1−)
potassium di**carbonylcyano**nitrosylcobaltate(0)

[CoH(N₂){(C₆H₅)₃P}₃]
(dinitrogen**)h**ydridotris(**t**riphenylphosphine)cobalt
(dinitrogen**)h**ydridotris(**t**riphenylphosphine)cobalt(I)

[NiCl₃(H₂O){N(CH₂CH₂)₃NCH₃}]
aquatri**c**hloro{1-**m**ethyl-4-aza-1-azoniabicyclo[2.2.2]octane}nickel
aquatri**c**hloro{1-**m**ethyl-4-aza-1-azoniabicyclo[2.2.2]octane}nickel(II)

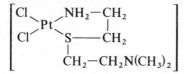

di**c**hloro[N,N-**d**imethyl-2,2′-thiobis(ethylamine)-S,N′]platinum
di**c**hloro[N,N-**d**imethyl-2,2′-thiobis(ethylamine)-S,N′]platinum(II)

The 1970 IUPAC rules give much further valuable information on how to name the extremely complicated structures that arise in this ever-growing class of compound. It cannot be fully reproduced in this introductory text, but the following examples give an idea of its range.

(i) Abbreviations

A valuable list of abbreviations recommended for common ligands is reproduced in *Table 2.5*.

(ii) π-Complexes

In some circumstances it suffices to give only the stoichiometric composition of π-complexes, e.g.

$$[PtCl_2(C_2H_4)(NH_3)]$$
amminedichloroethyleneplatinum

When all atoms in a chain or ring are bound to the central atom, structure may be indicated by use of the Greek letter η, which may be read as eta or hapto (from the Greek to fasten). The preceding compound would then be named:

amminedichloro(η-ethylene)platinum
or amminedichloro(η-ethylene)platinum(II)

A further selection of examples may be instructive:

[Cr(C₆H₆)₂]
bis(η-benzene)chromium
bis(η-benzene)chromium(0)

[ReH(C$_5$H$_5$)$_2$]
bis(η-**c**yclopentadienyl)**h**ydridorhenium
bis(η-**c**yclopentadienyl)**h**ydridorhenium(III)

[Co(C$_5$H$_5$)(C$_5$H$_6$)]
(η-cyclopentadien**e**)(η-cyclopentadien**yl**)cobalt
(η-cyclopentadien**e**)(η-cyclopentadien**yl**)cobalt(I)

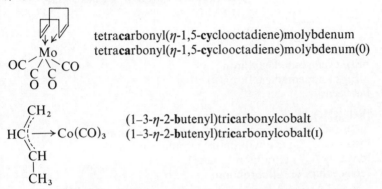

tetra**c**arbonyl(η-1,5-**c**yclooctadiene)molybdenum
tetra**c**arbonyl(η-1,5-**c**yclooctadiene)molybdenum(0)

(1–3-η-2-**b**utenyl)tri**c**arbonylcobalt
(1–3-η-2-**b**utenyl)tri**c**arbonylcobalt(I)

Table 2.5 ABBREVIATIONS FOR SOME COMMON LIGANDS

Hacac	acetylacetone, 2,4-pentanedione, CH$_3$COCH$_2$COCH$_3$
acac	acetylacetonato
Hbg	biguanide, H$_2$NC(NH)NHC(NH)NH$_2$
H$_2$dmg	dimethylglyoxime, 2,3-butanedione dioxime, CH$_3$C(=NOH)C(=NOH)CH$_3$
Hdmg	dimethylglyoximato(1–)
dmg	dimethylglyoximato(2–)
H$_4$edta	ethylenediaminetetraacetic acid, (HOOCCH$_2$)$_2$NCH$_2$CH$_2$N(CH$_2$COOH)$_2$
H$_2$ox	oxalic acid, HOOC—COOH
bpy	2,2'-bipyridine or 2,2'-bipyridyl,
diars	*o*-phenylenebis(dimethylarsine), (CH$_3$)$_2$AsC$_6$H$_4$As(CH$_3$)$_2$
dien	diethylenetriamine, H$_2$NCH$_2$CH$_2$NHCH$_2$CH$_2$NH$_2$
diphos	ethylenebis(diphenylphosphine), Ph$_2$PCH$_2$CH$_2$PPh$_2$
en	ethylenediamine, H$_2$NCH$_2$CH$_2$NH$_2$
phen	1,10-phenanthroline,
pn	propylenediamine, H$_2$NCH(CH$_3$)CH$_2$NH$_2$
py	pyridine
tren	2,2',2''-triaminotriethylamine, (H$_2$NCH$_2$CH$_2$)$_3$N
trien	triethylenetetraamine, (H$_2$NCH$_2$CH$_2$NHCH$_2$)$_2$
ur	urea, (H$_2$N)$_2$CO

tri**car**bonyl(1–4-*η*-**cy**clooctatetraene)iron

Fe(CO)$_3$

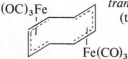

(OC)$_3$Fe

trans-*μ*-(1–4-*η*:5–8-*η*-cyclooctatetraene)-bis-(tricarbonyliron) (for *μ* see next section)

Fe(CO)$_3$

Fe(C$_5$H$_5$)$_2$
bis(*η*-cyclopentadienyl)iron
bis(*η*-cyclopentadienyl)iron(ii)
ferrocene

[Fe(C$_5$H$_5$)$_2$] [BF$_4$]
bis(*η*-cyclopentadienyl)iron(1+) tetrafluoroborate
bis(*η*-cyclopentadienyl)iron(iii) tetrafluoroborate
ferrocene(1+) tetrafluoroborate(1−)
ferrocenium tetrafluoroborate

Fe

O   2,4-(1,1′-ferrocenediyl)cyclopentanone

Ferrocene may be used as a parent, but multiplication of -ocene names is not favored by IUPAC.

(iii) Bridging groups

These are designated by the Greek letter *μ* (mu) and are then followed by a hyphen (this hyphen is desirable to differentiate, e.g. *μ*-dichloro-from a non-bridging dichloro occurring in the same complex).

[(NH$_3$)$_5$Cr−OH−Cr(NH$_3$)$_5$]Cl$_5$
μ-hydroxo-bis(pentaamminechromium)(5+) chloride
μ-hydroxo-bis[pentaamminechromium(iii)] chloride

[(CO)$_3$Fe(CO)$_3$Fe(CO)$_3$]
tri-*μ*-carbonyl-bis(tricarbonyliron)

[Br$_2$Pt(SMe$_2$)$_2$PtBr$_2$]
bis(*μ*-dimethyl sulfide)-bis[dibromoplatinum(ii)]

[(CO)$_2$Ni(Me$_2$PCH$_2$CH$_2$PMe$_2$)$_2$Ni(CO)$_2$]
bis[*μ*-ethylenebis(dimethylphosphine)]-bis(dicarbonylnickel)

$$\left[(H_3N)_3Co\underset{\underset{O}{\overset{|}{ON}}}{\overset{\overset{OH}{|}}{-}OH-}Co(NH_3)_3 \right]^{3+}$$

hexa**a**mmine-di-μ-**h**ydroxo-μ-**n**itrito(O,N)-dicobalt(3+) ion
hexa**a**mmine-di-μ-**h**ydroxo-μ-**n**itrito(O,N)-dicobalt(III) ion

$[Be_4O(CH_3COO)_6]$
hexa-μ-**a**cetato-(O,O')-μ_4-**o**xo-tetraberyllium
hexa-μ-**a**cetato-(O,O')-μ_4-**o**xo-tetraberyllium(II)

$[Cr_3O(CH_3COO)_6]Cl$
hexa-μ-**a**cetato-(O,O')-μ_3-**o**xo-trichromium(1+) chloride
hexa-μ-**a**cetato-(O,O')-μ_3-**o**xo-trichromium(III) chloride

(iv) Extended structures

Many compounds of extended structure exist; their nature is identified by a prefix *catena-*, as in:

$[Cs]_n\ [\cdot\cdot CuCl_2{-}Cl{-}CuCl_2{-}Cl{-}CuCl_2{-}Cl\cdot\cdot]^{n-}$

   cesium *catena*-μ-chloro-dichlorocuprate(II)

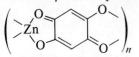

catena-di-μ-chloro-palladium

$$\left(\overset{>}{}Zn\underset{O}{\overset{O}{\diagdown\diagup}}\underset{}{\overset{}{\bigcirc}}\underset{O}{\overset{O}{\diagup\diagdown}} \right)_n$$

catena-μ-[2,5-dioxido-p-benzoquinone(2−)-$O,O':O'',O'''$]-zinc
catena-μ-[2,5-dioxido-p-benzoquinone(2−)-$O,O':O'',O'''$]-zinc(II)

(v) Di- and poly-nuclear compounds

When there is no bridging group, symmetrical di- and poly-nuclear compounds are named by use of bis-, etc., but for unsymmetrical compounds one component is treated as substituted by the others, as in:

$[Br_4Re{-}ReBr_4]^{2-}$
bis(tetrabromorhenate)(2−)
bis[tetrabromorhenate(III)]

[(C₆H₅)₃AsAuMn(CO)₅]
pentacarbonyl[(triphenylarsine)aurio]manganese

When there are also bridging groups, the metal–metal bond is indicated at the end of the name, e.g.

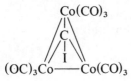

μ_3-iodomethylidyne-*cyclo*-tris(tricarbonylcobalt)(3 *Co—Co*)

Os₃(CO)₁₂
cyclo-tris(tetracarbonylosmium)(3 *Os—Os*)

An alternative name for the last compound is supplied in the next section.

(vi) Homoatomic aggregates (clusters)

The geometrical shape of the cluster is indicated by an abbreviated affix of geometrical significance, e.g. *triangulo-*, *quadro-*, *tetrahedro-*, *octahedro-*, *dodecahedro-*, etc. Examples are the following, but this type of geometrical labeling seems likely to be overwhelmed as examples proliferate.

Os₃(CO)₁₂
dodecacarbonyl-*triangulo*-triosmium

B₄Cl₄
tetrachloro-*tetrahedro*-tetraboron

[Mo₆Cl₈Cl₆]²⁻
octa-μ_3-chloro-hexachloro-*octahedro*-hexamolybdate(2—) ion
octa-μ_3-chloro-hexachloro-*octahedro*-hexamolybdate(ɪɪ) ion

(vii) Isomerism

Prefixes *cis-* and *trans-* are used for 4-planar and 6-octahedral complexes, supplemented by *fac-* (facial) and *mer-* (meridional), as illustrated.

In many cases locants are needed to distinguish isomers; the choice of locants is based in the IUPAC rules on locating planes of atoms perpendicular to a major axis and is described in elaborate rules. For octahedral complexes the method described in the

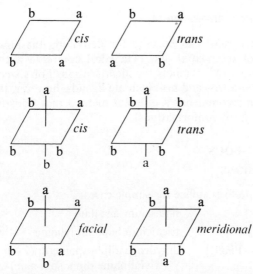

sequence-rule paper[6] is an alternative but does not have approval of the IUPAC Commission for the Nomenclature of Inorganic Chemistry.

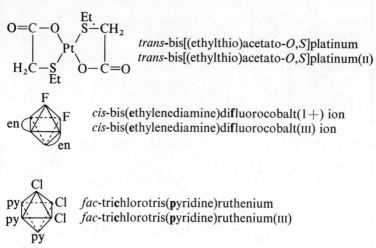

trans-bis[(ethylthio)acetato-*O*,*S*]platinum
trans-bis[(ethylthio)acetato-*O*,*S*]platinum(II)

cis-bis(**e**thylenediamine)di**f**luorocobalt(1+) ion
cis-bis(**e**thylenediamine)di**f**luorocobalt(III) ion

fac-tri**c**hlorotris(**p**yridine)ruthenium
fac-tri**c**hlorotris(**p**yridine)ruthenium(III)

mer-tri**c**hlorotris(**p**yridine)ruthenium
mer-tri**c**hlorotris(**p**yridine)ruthenium(III)

32   *Inorganic*

(viii) Absolute configuration

The sequence rule[6] can be used for describing the absolute configuration of tetrahedral and octahedral complexes; it prescribes (R), (S) or (P), (M) symbols for identification. For six-coordinated complexes with tris- and bis-bidentate ligands, however, the IUPAC Commission recommends a different method that leads to symbols Δ, Λ (or δ, λ for conformations).

MISCELLANEOUS

ISOPOLYANIONS

Numerical prefixes suffice for simple cases:

$K_2S_2O_7$	potassium disulfate
$Ca_3Mo_7O_{24}$	tricalcium heptamolybdate
$[O_2HP{-}O{-}PHO_2]^{2-}$	dihydrogendiphosphate(III)(2−) (trivial name diphosphonate)
$[O_2HP{-}O{-}PO_3H]^{2-}$	dihydrogendiphosphate(III,V)(2−)

Prefixes *cyclo-*, *catena-* or μ- are used, together with Stock numbers and charges, when it is felt to be desirable. Some examples are:

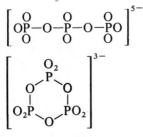

triphosphate

cyclo-triphosphate

$[O(PO_3)_n]^{(n+2)-}$ *catena*-polyphosphate

1-amidotriphosphate(4−)

1,2-μ-imidotetraphosphate(6−)

HETEROPOLYANIONS

Linear heteropolyanions are named by treating the terminal component that comes first in alphabetical order as having the others as ligands. If the terminal units are the same, choice rests on the second from the end, and so on.

$[O_3P—O—SO_3]^{3-}$ phosphatosulfate(3—)

$[O_3Cr—O—AsO_2—O—PO_3]^{4-}$ (chromatoarsenato)phosphate(4—)

Cyclic heteropolyanions receive the prefix *cyclo-*; alphabetical order decides the starting point and direction of citation, as in:

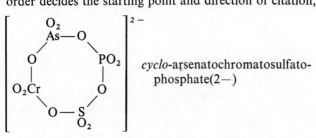

cyclo-arsenatochromatosulfato-
    phosphate(2—)

For a polyanion with an obvious central atom, the other components are named in alphabetical order as ligands, e.g.

$$\left[\begin{array}{c} O \\ O_3CrOPOAsO_3 \\ O \\ SO_3 \end{array} \right]^{4-}$$

(arsenato)(chromato)(sulfato)phosphate(4—)

Condensed heteropolyanions are named by designating the octahedral atoms surrounding the neutral atoms by prefixes such as wolframo- (or tungsto-), molybdo-, etc. Examples are:

$[PW_{12}O_{40}]^{3-}$ dodecawolframophosphate(3—)
                          or 12-wolframophosphate(3—)

$[Co^{II}Co^{III}W_{12}O_{42}]^{7-}$ dodecawolframocobalt(II)cobalt(III)ate

$Li_3H[SiW_{12}O_{40}]\cdot24H_2O$ trilithium hydrogen dodecawolframo-
                          silicate-24-water

ADDITION COMPOUNDS

Molecular compounds, solvates and clathrates are, according to the 1970 IUPAC rules, to be designated by connecting the names of the components by hyphens (or en rules) and stating the molar proportions by Arabic numerals separated by a solidus and in parentheses after the name. Boron compounds and water are always cited last in that order. Other molecules are cited in order of increasing number; any that occur in equal numbers are cited in alphabetical order. In formulas, dots at the midway position replace the hyphens. For example:

$Na_2CO_3 \cdot 10H_2O$	sodium carbonate–water(1/10) or sodium carbonate decahydrate
$NH_3 \cdot BF_3$	ammonia–boron trifluoride(1/1)
$BF_3 \cdot 2H_2O$	boron trifluoride–water(1/2)
$8CHCl_3 \cdot 16H_2S \cdot 136H_2O$	chloroform–hydrogen sulfide–water–(8/16/136)

NON-STOICHIOMETRIC CRYSTALS.

Crystalline phases not conforming to strict stiochiometry are discussed at some length in the 1970 IUPAC rules (Section 9), where designations for phases of variable composition (berthollides), vacant and interstitial sites (including Schottky and Frenkel defects), surface sites, and doped materials are dealt with. A symbol □ is used in many such cases.

BORON COMPOUNDS

Considerable attention is paid in the 1970 IUPAC rules to this rapidly growing branch of chemistry, and the main principles will be noted here.

Names of boron hydrides are patterned as BH_4 borane, B_2H_6 diborane(6), B_6H_{10} hexaborane(10), etc., and the numbering of the boron atoms is carefully laid down.

For polyboranes, those having closed structures receive a prefix

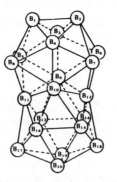 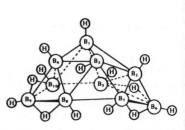

$B_{20}H_{16}$ *closo*-icosaborane(16) skeleton $B_{10}H_{14}$ *nido*-decaborane(14)

closo-, and those with structures very nearly a closed structure receive a prefix *nido-* (Latin *nidus*, nest). These two instructions are obviously indefinite but the illustrations on p. 34 will show the intention. Note incidentally the use of icosa (not eicosa) for boron chemistry. These rules also suggest commo where organic chemists use spiro, and make use of *endo* and *exo* in loosely defined ways. Such nomenclature must be treated with caution.

Organic ring-fusion principles can be used to designate detailed ring structure, with the names of the simple boranes as components. Derivatives are named by substitution (amino, etc., prefixes), and bridging groups by μ.

Radical names are:

H_2B-	boryl	H_2BH_2BH-	diboran(6)yl
$HB\!\!<$	boranediyl	$-HBH_2BH-$	1,2-diborane(6)diyl
$B\!\!\ll$	boranetriyl	$H_2BH_2B\!\!<$	1,1-diborane(6)diyl

$$\begin{array}{c} H \\ \diagup \;\; \diagdown \\ H_2B \qquad BH_2 \\ \diagdown \;\; \diagup \end{array} \quad \text{(1–2)diboranyl etc.}$$

The customary carborane nomenclature, e.g. $B_{10}C_2H_{12}$ dicarba-dodecaborane(12), is approved.

Some further detail is given in a set of Tentative rules[7] that at the time of writing await final approval or amendment.

PREFIXES AND AFFIXES IN INORGANIC NOMENCLATURE

The useful list in *Table 2.6* is reproduced from the IUPAC 1970 rules.

Table 2.6   PREFIXES OR AFFIXES USED IN INORGANIC NOMENCLATURE

Multiplying affixes (a)	mono, di, tri, tetra, penta, hexa, hepta, octa, nona (ennea), deca, undeca (hendeca), dodeca, etc., used by direct joining without hyphens
(b)	bis, tris, tetrakis, pentakis, etc., used by direct joining without hyphens, but usually with enclosing marks around each whole expression to which the prefix applies
Structural affixes	italicized and separated from the rest of the name by hyphens
antiprismo	eight atoms bound into a rectangular antiprism

(contd. on p. 36)

Table 2.6 (contd.)

asym	asymmetrical
catena	a chain structure; often used to designate linear polymeric substances
cis	two groups occupying adjacent positions; sometimes used in the sense of *fac*
closo	a cage or closed structure, especially a boron skeleton that is a polyhedron having all triangular faces
cyclo	a ring structure*
dodecahedro	eight atoms bound into a dodecahedron with triangular faces
fac	three groups occupying the corners of the same face of an octahedron
hexahedro	eight atoms bound into a hexahedron (e.g. cube)
hexaprismo	twelve atoms bound into a hexagonal prism
icosahedro	twelve atoms bound into a triangular icosahedron
mer	meridional; three groups on an octahedron in such a relationship that one is *cis* to the two others which are themselves *trans*
nido	a nest-like structure, especially a boron skeleton that is very close to a closed or *closo* structure
octahedro	six atoms bound into an octahedron
pentaprismo	ten atoms bound into a pentagonal prism
quadro	four atoms bound into a quadrangle (e.g. square)
sym	symmetrical
tetrahedro	four atoms bound into a tetrahedron
trans	two groups directly across a central atom from each other; i.e. in the polar positions on a sphere
triangulo	three atoms bound into a triangle
triprismo	six atoms bound into a triangular prism
η (eta or hapto)	signifies that two or more contiguous atoms of a group are attached to a metal
μ (mu)	signifies that the group so designated bridges two or more centers of coordination
σ (sigma)	signifies that one atom of the group is attached to a metal

* A prefix *cyclo-* here is used as a modifier indicating structure and hence is italicized. In organic nomenclature, cyclo- is considered to be part of the parent name since it changes the molecular formula and therefore is not italicized.

REFERENCES

1. See *J. Chem. Soc.*, 1404 (1940)
2. *IUPAC Compt. Rend. 17th Conf.*, 98–119 (1953)
3. *IUPAC Nomenclature of Inorganic Chemistry, 1957 Rules*, Butterworths, London (1959)
4. *IUPAC Compt. Rend. 23rd Conf.*, 181–187 (1965)
5. *IUPAC Nomenclature of Inorganic Chemistry*, 2nd edn., *Definitive Rules 1970*, Butterworths, London (1971) [reprint from *Pure and Appl. Chem.*, **28** (1971)]
6. Cahn, R. S., Ingold, Sir C. and Prelog, V., *Angew. Chem., Internat. Ed.*, **5**, 385, 511 (1966)
7. *IUPAC Information Bulletin, Appendices on Tentative Nomenclature, Symbols, Units, and Standards*, No. 8 (September 1970); also printed in *Inorg. Chem.*, **7**, 1948 (1968)

3 Organic: General

INTRODUCTION

Nomenclature rules published by the International Union of Pure and Applied Chemistry (IUPAC) are accepted by chemists of most countries and thus form the international authority for organic as for inorganic chemistry*.

The IUPAC Organic Nomenclature Rules A, B and C, 1969[2], which supersede earlier versions[3], cover most of general organic chemistry but hardly any of the specialized fields. Organic derivatives of P, As, Sb and Bi, and organometallic chemistry other than that for coordination complexes (see pp. 14, 24) are being considered jointly by the IUPAC Commissions on the Nomenclature of Organic and of Inorganic Chemistry, but at the time of writing nomenclature rules have not yet been published. A number of fields of major interest in both biochemistry and organic chemistry have been treated jointly by Commissions of IUPAC and IUB (International Union of Biochemistry), and several valuable sets of specialist rules have resulted (see Chapter 7).

However, anyone attempting to learn nomenclature by reading these rules A, B and C must find them often complex, arbitrary, contradictory, or vague (because of alternatives). These bad features arose from a wish by IUPAC (and all chemists) to retain as much of customary nomenclature as is still useful, because large sections of that customary usage are antique (e.g. acids and their derivatives are named by a dualistic principle reminiscent of the earliest chemists, and amines are often named on the mid-nineteenth century theory of types), and because chemists are loth to abandon a host of abbreviated names. Add to these the many thousands of purely trivial names, some old, some new, and it must surely be agreed that the time is ripe for a fully logical nomenclature, simple to understand

---

* Early attempts to reach agreed organic chemical nomenclature have been recorded by Verkade[1] in a series of papers containing much hitherto little-known or unpublished material.

and either adaptable for or derived from computer usage (cf. p. 1). But, until that is available, current IUPAC nomenclature needs to be mastered, not least because it is used with only minor exceptions in *Chemical Abstracts* indexes. In the account that follows here an attempt is made to set out not so much the formal rules as the most important principles, together with indications as to where information on the more recondite matters can be found. It will be most convenient if, before proceeding to the principles of nomenclature themselves, we first define certain technical terms and the conventions used in committing chemical names to paper.

TECHNICAL TERMS

Systematic name

A name composed wholly of specially coined or selected syllables: hexane, thiazole.

Trivial name

A name no part of which is used in a systematic sense: xanthophyll, furan.

Semisystematic = semitrivial name

A name of which only a part is used in a systematic sense: meth*ane*, but*ene*, calcifer*ol*. Most names in organic chemistry belong to this class; many are often spoken of loosely and incorrectly as simply 'trivial'.

Parent

We may think of benzene as the chemical parent of nitrobenzene because the latter is made from the former; and benzene is the parent name of nitrobenzene because the latter is derived from the former by a prescribed variation. However, parent names do not always reflect chemical parentage: ethane is hardly the chemical parent of ethanol, though ethane is the parent name of ethanol. Often there is a chain of parentage—hexane, 3-methylhexane, 3-(chloromethyl)hexane—or multiple parentage—benzanthracene from benzene and anthracene. It is important to note that a parent name may be systematic (e.g. hexane), trivial (e.g. furan) or semitrivial (e.g. methane).

Group or radical

A group of atoms common to a number of compounds (as in inorganic chemistry); e.g. CH_3, CH_2, OH, NO_2, CO, COOH. Such groups are often called radicals in nomenclature parlance. Free radicals are always given the designation 'free' in nomenclature work.

Function, functional group

The terms 'function' and 'functional group' entered English chemical nomenclature in IUPAC rules by translation from the 1930 French version (fonction, fonctionnel) without definition. A functional group is a group of atoms defining the 'function' or mode of activity of a compound. An alcohol owes its alcoholic properties to the functional group –OH; here the functional group, hydroxyl, is the same as the chemical group of the same name. A ketone owes its ketonic properties to the oxygen atom that is doubly bound (to carbon); the ketonic function is O=(C) (without the carbon), and this is not the same as the ketonic group $O=C\diagdown$. Similarly the carboxylic function is $\underset{HO\diagup}{O\diagdown}(C)$; $\underset{HO\diagup}{O\diagdown}C-$ is the carboxyl group. In chemical writing we often find 'function', 'group' and 'radical' used interchangeably; but they are not interchangeable; usually group or radical expresses the meaning intended. Incidentally, the division of substituents into 'functional' and 'non-functional' in Beilstein's *Handbuch* is also incorrect, or at best a special usage.

Principal group

Much confusion has been caused also by wording of the 1930 IUPAC rules where 'function' appeared sometimes to denote a group named as suffix, in distinction from one named as prefix. Now this distinction is an important one and the 1965 IUPAC rules introduced the term 'principal group' to denote the group named as suffix. It will be seen below that for any given compound only one kind of group may be named as a suffix and that its choice is regulated by definite rules. Use of 'principal group' in this sense allows 'function' to be used unambiguously in its original connotation of activity, with 'functional group' available to denote the group responsible for that activity.

Locant

This useful word denotes the numeral or letter that indicates the position of an atom or group in a molecule.

CONVENTIONS

Conventions concerning the exact way in which chemical names should be written are an annoyance because so many of them are not emphasized in speech and because they vary from country to country and in some cases from journal to journal within one country, partly for linguistic reasons and partly from personal preferences. Some of these differences chemists take in their stride, but others must be assiduously learnt: for instance, it is important to know the difference between 1,2,5,6- and 1,2:5,6- and that italicized prefixes are not alphabeticized in most chemical indexes, so that isobutane appears under 'i' whereas *sec*-butane appears under 'b'.

Particularly distressing are the differences between *Chemical Abstracts* and the (British) Chemical Society. In previous editions of this book the British conventions were used. In the present edition, since computer output from UKCIS and other organizations is based almost wholly on American compilations, and since American publications are much the more numerous, it has seemed advisable to use American conventions. Differences between the two national habits are noted where appropriate.

Numerals

A numeral is a locant when it indicates the position of a substituent or bond in a structure. Two or more locants denoting the positions of two or more identical substituents are separated by commas, as in 1,2,5,6-tetrabromocyclohexane. When two or more operations each require two (or occasionally more) locants, the pairs (or triplets, etc.) of numerals are separated by colons, as in 1,2:5,6-di-*O*-isopropylideneglucitol. There is no space after the comma or colon. Single numerals, or such sets of numerals, are joined by hyphens to the following and any preceding parts of the name; however, in a few special cases a single letter is also part of the locant and then follows the numeral directly, between it and the hyphen or comma, e.g. 3α- or (2*H*)-.

Exceptionally, in bicyclo, tricyclo, etc., names (p.77) and in certain spiro names (p. 79) numerals are separated by full stops (periods), again without spaces.

Series of numerals are arranged in ascending order. An unprimed locant is lower than the same locant primed, a singly primed lower than a doubly primed, and any primed locant lower than a higher unprimed locant; e.g. an order would be 2,2,2,',3,3',3'',4. Special significance attaches to departure from this order in specific cases, e.g. bicyclo and spiro names (pp. 77, 79).

Locants are placed as early in a name as does not cause confusion. Simple examples are 2-chlorohexane, 2-chloro-3-methylhexane and 2,3-dichlorohexane.

American practice extends this to locants for suffixes in cases such as 2-hexanol, 2,3-hexanediol, 2-phenanthrenecarboxylic acid and 2-hexene. But this can be done only for one *type* of such ending; if there is more than one, the locant appearing first in the name is moved to the left and the others directly precede their suffixes, e.g. 2-hexen-4-yne, 3-hexen-5-yn-2-ol. That practice is followed in this edition. British custom is generally to place the locant always immediately in front of its suffix, as in hex-2-ene and hex-3-en-5-yn-2-ol, but most chemists in other countries dislike this as splitting spoken words unnecessarily.

There are also cases where no locant for a suffix can be moved to the left, namely when some other locant must by a special rule be attached at that place; examples such as 5α-cholestan-3-one and 2*H*-pyran-3-ol illustrate this.

After all this, it should be noted that quite different rules apply to positions and punctuation of locants in many non-English languages (cf. p. 46).

Letter locants

Single-letter locants are also common, often italicized atomic symbols; if for identical groups, such locants are arranged in alphabetical order, Latin before Greek, whether capital or lower case. Arrangement for primed letters follows the same principles as for primed numerals. Mixtures such as 2,*N*,α-trimethyl occasionally occur.

Multiplying prefixes

The normal multiplying prefixes used in organic chemistry are listed in *Table 3.1*.

Table 3.1   MULTIPLYING PREFIXES NORMALLY USED IN ORGANIC CHEMISTRY

1	mono	11 undeca*	
2	di	12 dodeca	
3	tri	13 trideca	30 triaconta
4	tetra		31 hentriaconta*
5	penta		32 dotriaconta
6	hexa	20 eicosa†	
7	hepta	21 heneicosa*	40 tetraconta
8	octa	22 docosa	
9	nona*	23 tricosa	100 hecta
10	deca	24 tetracosa	132 dotriacontahecta

* nona and un are from Latin; ennea and hen from Greek.
† Contrast icosa for inorganic chemistry (p. 35).

   Multiplicative prefixes bis-, tris-, tetrakis-, etc. are used in America for all complex expressions. The commonest is for compound radicals, e.g. bis(dimethylamino), tris(2-chloroethyl), tetrakis(*p*-tolyl), etc.; and this is extended to other situations where parentheses aid clarity, as in bis(diazo), tris(hydrogen phthalate) and tris(methylene) (to avoid confusion with trimethylene $-CH_2CH_2CH_2-$).
   British practice has been more permissive. Bis, etc. have been used when the next syllable was numerical, as in bis-2,2,2-trichloroethyl, parentheses being considered unnecessary; and di-(2-chloroethyl) has been held in Britain to be as unambiguous with the parentheses as with a bis prefix.
   It will also be found that the first hyphen that in Britain would be in tris-(2-chloroethyl) and the like is considered in America to be unnecessary when a parenthesis follows; the American custom would be tris(2-chloroethyl); strict grammar requires that parentheses should not alter punctuation in any way; the Americans prefer their abbreviated usage.

Parentheses, brackets, etc.

As just noted, in *Chemical Abstracts*, parentheses ( ) are used around all compound radical names to avoid any possible confusion, e.g. 2-(bromomethyl), (phenylazo), and even 2-(*p*-tolyl), 2-(2-naphthyl), etc. In more complex cases, square brackets [] and if necessary

braces {} are added, e.g. 1-[2-(diethylamino)ethyl]naphthalene, or
1-{3-[2-(diethylamino)ethyl]pentyl}carbazole. British practice has
been less rigid, allowing chemical sense to dictate when parentheses
are needed, with primes used in some cases so as to avoid the need
for parentheses, e.g. 1-2'-chloroethyl rather than 1-(2-chloroethyl).
Chemical sense is also used in American practice when ambiguity
might otherwise arise: for instance, CH_3PHCl is chloromethyl-
phosphine, but PH_2CH_2Cl is (chloromethyl)phosphine; so also
glucose 6-(dihydrogen phosphate); and abbreviated names such as
methoxy are not placed within parentheses whereas unabbreviated
(heptyloxy), etc. are.

Many other special uses for enclosing marks will be found in the
following pages.

Italics

Since italicized syllables are not 'counted' in alphabetization, their
use in the following pages should be noted. The purpose of italics
is almost always simply to ensure that index entries occur in the
most useful places, keeping isomers and related compounds in
suitable groups. This being the object, the fact that no difference is
made in speech becomes irrelevant.

A lower-case italic prefix at the beginning of a line or after a full
stop (period) is not given an initial capital—the next letter becomes
a capital, as in *m*-Xylene or *trans*-2-Butene.

Elision of vowels

Vowels are elided in *Chemical Abstracts* for the following cases *only*:
(i) before a functional suffix that begins with a vowel, e.g. 2-hexanone
(not 2-hexaneone); (ii) between 'a' of a numerical prefix and a
functional suffix that begins with a vowel, e.g. benzenehexol (not
benzenehexaol); (iii) in Hantzsch–Widman names (see p. 81), e.g.
oxazole (not oxaazole); or (iv) when cyclobuta or cyclopropa is
attached to a further part of a ring name that begins with a vowel,
e.g. cyclobutindene. Note that (ii) is not applied in inorganic chemistry by IUPAC (see p. 9).

Addition of vowels

An 'o' is occasionally inserted for euphony between a consonant
ending one part of a name and a consonant (notably 'h') beginning
the next part, e.g. acetohydroxamic acid (not acethydroxamic).

Formulas

In simple groups the central atom is written first, e.g. CH_3, CH_2, NO_2, $N(CH_3)_2$, $NH(CH_3)$, etc.; CH_3CH_2 may be shortened to C_2H_5, and similarly for other alkyl groups. When such a group appears on the left of a formula the order may be reversed so as to emphasize the bonded atoms, as in $O_2N-CH_2CH_2-OH$ or $ON-O-CH_2CH_2OH$.

In British journals successive groups in a linear formula are often separated by points at the midway position, as in $CH_3 \cdot CH_2 \cdot CH_2 \cdot OH$. But it is widely held elsewhere that such points should be restricted to denoting free radicals or addition compounds. Such 'British' midway points are never used in American practice in linear organic formulas; to emphasize the existence of a bond, a hyphen or 'rule' is inserted, as in $O_2N-CH_2CH_2OH$.

Parentheses are used freely in linear formulas, for example $(CH_3)_2NCH_2CH(CH_3)C(CH_3)_2OH$.

Certain group symbols are very frequent in British publications but are rarer elsewhere and do not form part of official American or IUPAC practice; the commonest of these are:

Me	CH_3-	Bu^i	$(CH_3)_2CHCH_2-$
Et	CH_3CH_2-	Bu^s	$CH_3CH_2CH(CH_3)-$
Pr^n	$CH_3CH_2CH_2-$	Bu^t	$(CH_3)_3C-$
Pr^i	$(CH_3)_2CH-$	Ac	CH_3CO-
Bu^n	$CH_3CH_2CH_2CH_2-$	Ph	C_6H_5-
	Bz C_6H_5CO-	(benzyl is $PhCH_2-$)	

In some specialist nomenclature, particularly in biochemistry and for coordination compounds, special symbols are essential for brevity; and indeed any symbol may be used if explained in the text.

R is widely used as a generic symbol for a group. A series of such should be written R, R′, R″..., or R^1, R^2, R^3...., but not R_1, R_2, R_3 (which denote 1, 2 or 3 identical R groups at one position in a structure).

Hyphenation of chemical words

Hyphenation of chemical terms is much more common in British than in American practice. In the former a hyphen is inserted between two identical letters in a chemical name (e.g. tetra-acetate, methyl-lithium), to help in reading such words as co-ordination and un-ionized (not a union), to separate portions of partial names such as keto-ester, amino-acid, oxo-steroids, and when partial chemical names end in a voiced vowel or 'y', e.g. amino-derivative, thia-compound, methoxy-group (but methyl derivative, etc.). None of these hyphens is used in American practice: identical letters are simply run on (tetraacetate, etc.), and partial chemical names are followed by a space (amino acid, keto group). In all these respects the British practice is becoming more elastic.

OTHER LANGUAGES

By and large, chemical names are much easier than current text to translate from foreign languages, for they almost always bear at least some resemblance to one or other of the not too ancient systems current in the UK or USA; it is the long-known compounds that are the usual exceptions to this emollient statement, for example, the common metals, the common gases, and a few organic compounds such as acetic and formic acid. Most troublesome are the resemblances that hide differences, and the following notes may be useful.

In German the main trap is that a terminal 'e' on a word denotes a plural (with few exceptions). 'Anilin' is aniline, 'Nitroaniline' are nitroanilines; 'Phenol' is phenol, 'Phenole' are phenols; thiazole becomes 'Thiazol', and thiazoles become 'Thiazole'. Thus the distinction of basic -ine from non-specific -in (tannin, protein, anthocyanin, vitamin, etc.) and of alcoholic or phenolic -ol from -ole (denoting a five-membered aromatic ring) cannot be used in German, and the true significance of the 'e' in German may be overlooked by a hasty reader.

The ability, in German, to pile nouns on one another shines also through their nomenclature, but, for example, acetessigsaureethylesterdinitrophenylhydrazone merely requires careful dissection. More subtle are the results where in German there may be more hyphens than in English, as in 'Chloro-bromo-phenylnaphtalin', or fewer hyphens as in 'Chloro- und Bromonaphtaline'. Perhaps too a special note may be needed that the German for benzene is 'Benzol'.

In general, the influence of Beilstein/Stelzner nomenclature in German literature is still profound, though IUPAC Nomenclature is steadily gaining ground, principally because of its greater simplicity (or lesser complexity!) and wider applicability together with the many publications of the Chemical Abstracts Service.

The problem of terminal 'e' has a different result in French, namely that, for phonetic reasons, -in and -an endings for non-nitrogenous rings (p. 86) become -inne and -anne. But the features in French nomenclature that most strike the English-speaking reader are that locants follow the name of the structural feature concerned, and the division of names into small words, as in:

$$ClCH_2-CH-CH_2-CH-CH_2CH_2CH_3$$

$$CH_2Cl \quad CHClCH_3$$

Chloro-1 (chloro-1 éthyl)-4 chlorométhyl-2 heptane

There is also the formation of salt and ester names as, for example, 'acetate de sodium', 'malonate de diéthyle', etc.

An excellent exposition of French organic nomenclature, reproducing the latest IUPAC rules (with occasional modifications) and including a brief historical introduction and very valuable comments on individual sections and rules, has been published by Lozac'h[4]; it is a much more sophisticated and detailed treatment than is given in the present book and is, to the writer's knowledge, the only one of its kind in any language.

The Russian chemical literature now presents fewer problems than it used to, because most of it is also published in English translations; but the reader should be warned that Russian nomenclature seems to be left to the author's whims and is often modelled on Beilstein rather than IUPAC—a feature that can be particularly troublesome in their extensive organophosphorus papers.

In other countries, notably Scandinavia, Japan, Holland and Belgium, the tendency to publish in English increases continuously, then mainly in the style of *Chemical Abstracts* and IUPAC.

REFERENCES

1. Verkade, P. E., *Bull. Soc. Chim. France*, 1807 (1966); 4009 (1967); 1358 (1968); 3877, 4297 (1969); 2739 (1970); 1634, 4299 (1971)
2. *IUPAC Nomenclature of Organic Chemistry. Definitive Rules for: Section A. Hydrocarbons; Section B. Fundamental Heterocyclic Systems; Section C. Characteristic Groups Containing Carbon, Hydrogen, Oxygen, Nitrogen, Halogen, Sulfur, Selenium and/or Tellurium.* 1969. A,B 3rd edn.; C 2nd edn., Butterworths, London (1971)
3. Ref. 2: Sections A and B, 1st edn. 1958, 2nd edn. 1966. Section C, 1st edn. 1965 (also printed in *Pure and Appl. Chem.*, **11**, Nos. 1–2 (1965)
4. Lozac'h, N., *La Nomenclature en Chemie Organique*, Masson et Cie, Paris (1967) (Vol. 6 of *Collection de 'Monographies de Chimie Organique, Compléments au 'Traité de Chimie Organique'*, under the direction of A. Kirrmann, M.-M. Janot and G. Ourisson)

4 Organic: The Principles

TYPES OF NOMENCLATURE

There is a fundamental distinction between the use of trivial and of systematic names: trivial names refer to compounds, systematic names to structures, i.e. structural formulas. Trivial names are independent of structure; they can be, and often are, assigned before the structure is known; and when the structure is known, the one name embraces all dynamic variations due to tautomerism, etc. A systematic name, being derived from one formula, cannot, if it is accurately descriptive, apply to a tautomer thereof (though normally it covers changes due to resonance, hyperconjugation, etc.). Of the approximately three million organic compounds at present recorded, many thousands have trivial or semitrivial names, each of which, not being wholly logical, requires some feat of memory for recall of the relevant structure. No-one, of course, remembers more than a proportion of them; yet trivial names are inevitable when of long tradition or when the systematic name is too unwieldy for use—certainly for macromolecules such as proteins, nucleic acids, etc. The important thing is to avoid coining new ones just for the fun of it and in simple cases gradually to replace the old by the systematic.

In a semitrivial name an indication of partial structure is, by definition, built in (e.g. the alcoholic group in ethanol); it is essential that this be done correctly, e.g. that acidic phenols be not *named* acids, and that variations of the trivial name for derivatives be made in accordance with the rules for systematic nomenclature.

Owing to the formation, by covalent linkages, of chains and rings of carbon atoms, alone or with hetero atoms (i.e. atoms other than carbon), the structures of organic compounds differ fundamentally from those of many inorganic compounds. The two nomenclatures therefore use quite different methods. Organic nomenclature has, however, been of slow growth and in its present state no less than eight general principles and several specialized principles can be distinguished. The latter will be recorded at appropriate places. The former can be outlined at once as follows; in most cases they apply equally to trivial, semitrivial and systematic names.

(a) The basic principle is substitution, the replacement of hydrogen by an atom or group, e.g. of H by Cl (chlorination), by NO_2 (nitration), by CH_3 (methylation), even though this replacement may not be the method of synthesis.

The reverse also holds. Chlorination, for instance, is only replacement of H by Cl; the term should not be used for addition, as in: $C_6H_5CH{=}CH_2 + Cl_2 \rightarrow C_6H_5CHClCH_2Cl$ or the reaction

$$\text{>C–OH} \rightarrow \text{>C–Cl}.$$

Substitution is indicated by a suffix (ethane, ethanol) or a prefix (benzene, chlorobenzene), the loss of hydrogen not being stated.

(b) By substitutive nomenclature one reaches the name 2-naphthylacetic acid for $2\text{-}C_{10}H_7CH_2CO_2H$, substitution of the naphthyl radical into acetic acid. Such names are, however, not useful for indexing; the parent is acetic acid, and after inversion the index entry would be Acetic acid, 2-naphthyl-; but the compound is more suitably indexed under the larger part, naphthalene, and this can be achieved by juxtaposing the two names naphthalene and acetic acid and preceding them by the positional numeral. This yields 2-naphthaleneacetic acid (for indexing under N), the loss of *two* hydrogen atoms being assumed in this process (just as the loss of *one* is assumed in substitutive names). This is termed 'conjunctive nomenclature'. Further detail is given on pp. 60–61.

(c) What is superficially a mixture of substitutive and conjunctive nomenclature is the (universal) practice with prefixes for bivalent groups, such as –O– oxy, $\text{>C}{=}\text{O}$ carbonyl and –S– thio. Suppose we wish to treat CH_3OOC- as a substituent: we place the radical name methoxy next to the bivalent radical name carbonyl and obtain methoxycarbonyl $CH_3O{-}CO-$, *no* hydrogen being lost in the process; then we can substitute this group into, say, glycine NH_2CH_2COOH, with the usual loss of one hydrogen atom in the process, and in this way we obtain *N*-(methoxycarbonyl)glycine $CH_3OOCNHCH_2COOH$.

(d) For some classes old names indicating the function of the compounds survive, e.g. ethyl *alcohol*, diethyl *ether*, ethyl methyl *ketone*, acetic *acid*. The systematist would like these to disappear, for (except for acids) there are alternatives more in keeping with current practice; but such names die hard, if only because the chemist is, after all, interested mainly in the functions, i.e. modes of activity, of his compounds. This system has been termed radicofunctional nomenclature by IUPAC, because the functional class name (alcohol, ketone, etc.) is preceded by a radical name or names (ethyl, acetic, etc.).

(e) Replacement nomenclature, also called 'a' nomenclature: Sometimes replacing a hetero atom in a compound A by carbon would yield a substance B that is more readily recognized or more simply named. It is then possible to name the compound B and indicate the presence of the hetero atom by a prefix ending in 'a'. Thus, pyridine might be termed azabenzene. Actually this method is normally reserved for complex structures, such as 4-azacholestane, or for the presence of a multiplicity of hetero atoms. Its extension to aliphatic compounds is envisaged in IUPAC rules but ran into difficulties there; the only common linear use is an old variant, namely azomethine for $-N=$. It will be appreciated, nevertheless, that this method offers a possible future simplification of much organic nomenclature in general.

(f) A few additive reactions survive as a basis for names, e.g. styrene oxide. This is, in fact, one of the few additive names that are worth preserving, for the alternatives 1,2-epoxyethylbenzene and 1-phenyloxirane are less easily recognized by some chemists. Names such as ergosterol dibromide are also valuable as trivial names preserving the parent component and, above all, its stereochemistry. In general, however, additive nomenclature should be avoided: it forms no part of modern nomenclature. Of course, there is an exception: hydro for addition of hydrogen!

A different kind of additive nomenclature is used when an element increases its valence: e.g. pyridine gives pyridine 1-oxide.

(g) Subtraction: removal of atoms is indicated in a few cases, e.g. dehydro (loss of 2H), anhydro (loss of H_2O), nor (loss of CH_2), de (loss of a specified group; e.g. de-N-methyl).

(h) Many cyclic skeletons have trivial names; complex cases are handled by joining names of simpler components, by methods described in Chapter 5. There are also systems for such compounds where specific syllables have prescribed meanings for ring structure.

(i) A few chemical operations can be designated by specific affixes, e.g. lactone, seco-, -oside.

THE APPROACH TO A NAME

With such an array of possible procedures how does one set about naming a compound of known structure? Occasionally the chemistry under discussion is such that the expert may be justified in breaking the rules: but here let us forget such relatively rare occasions and confine ourselves to the classical methods. Then perhaps the obvious way is to look for a large recognizable unit—a ring structure or a long chain. Nevertheless, that would be wrong. The first thing to do is to seek out the substituent groups: OH, NH_2, COOH, SO_3H, OCH_3, $COOC_2H_5$, etc.; then from those present, the 'senior', so-called 'principal' group is found by means of *Table 4.2* (p. 52). That group, and that alone, is written as suffix, the others become prefixes. The principal group sets the whole pattern of nomenclature and numbering, and it is vital to fix that group before anything else is done. Two simple examples show this: $C_{10}H_7C_6H_4COOH$ is a naphthylbenzoic acid and not a carboxyphenylnaphthalene, whereas $C_6H_5C_{10}H_6COOH$ is a phenylnaphthoic acid; $NH_2CH_2CH_2OH$ is 2-aminoethanol, and not 2-hydroxyethylamine (because OH is senior to NH_2).

For the rest of the name one works back from the principal group. The possibilities are so various that the following outline will appear confusing, but read slowly it will be found to make sense.

(*A*) Suppose that there is only one type of principal group and that it is attached to an aliphatic chain. If there is also only one group of this type, choose the most unsaturated chain to which it is attached; if there are more than one of equal unsaturation—and this includes the case when there is no unsaturation—choose the longest. Or, if there are several groups of the same type, i.e. more than one principal group, attached to an aliphatic chain, then choose the straight chain containing the most of them; if there are two such chains, choose the more unsaturated, then the longer, as in the previous case. Next number the chain, with due regard to unsaturation and substituents, but giving the lowest available number(s) to the principal group(s), as explained later. Lastly, name the chain, add the suffix for the functional group(s), name the other substituents as prefixes and add them in alphabetical order to complete the name.

Or (*B*) suppose the principal group is attached to a ring. Then name the cyclic system, as explained later; number it, giving the lowest available numbers first to 'indicated' hydrogen (see p. 112) if any, then to the principal group, and finally to the other atoms and groups to be cited; and write down the name.

(*C*) If the principal group is attached to a chain and that in turn to a ring, the British author may treat the cyclic radical as a substituent into the chain; or conjunctive nomenclature (p. 60) may be used.

(*D*) A variant of these procedures is useful for compounds with the symmetrical structural pattern X–Y–X when each of the units X contains the same principal group for citation as suffix. Examples are

$$p\text{-HOOCCH}_2\text{–C}_6\text{H}_4\text{–CH}_2\text{COOH}$$

and $\quad \text{H}_2\text{NCH}_2\text{CH}_2\text{–O–CH}_2\text{CH}_2\text{NH}_2$

The principle is to cite the X groups with a prefix di- and to place in front of this the name of the bivalent group Y (i.e. phenylene and oxy in the two cases exemplified). This method is discussed in some detail on p. 64.

There are only a few steps in each case (*A–D*) and, if taken in the right order, they lead simply to the answer. Even though these are simplified directions and more complex situations are frequent, it is surprising how many of the compounds met in ordinary chemistry can be named merely by these simplified procedures. Of course, it is easier to write 'name the ring' than to do it, and tricky problems are often met. So in the following pages the procedures given will be expanded, and some of the problems discussed, but to reach them we must start again systematically at the beginning.

THE PRINCIPAL GROUP

Approaching a name again from the beginning, we must start with what we have just said is the first thing to do—selecting the principal group, i.e. selecting the type of group to be named as suffix.

But we must here at once note that there may not be any principal group in the compound; that will be the case if our compound is a hydrocarbon or heterocycle or has as substituents only groups that must be named as prefixes. The complex matter of naming hydrocarbons and heterocycles will require separate treatment (see Chapter 5). For the moment we shall remain with the groups.

As to groups that cannot be named as suffix, we find that there are fewer of these in substitutive than in the older radicofunctional nomenclature and, since substitutive nomenclature is much the more widely used as well as being the more modern, we shall start with that.

Principal groups in substitutive nomenclature

Table 4.1 (p. 52) lists the groups that, in substitutive nomenclature can be named only as prefixes, together with the names of those prefixes.

Table 4.2 lists, as classes, a far greater number of groups that *can* be named as suffix; any one of these, as sole substituent or accompanied only by group(s) listed in *Table 4.1*, is treated as the principal group, i.e. *must* be named as suffix. Examples are hexanoic acid, pentanal, 3-acenaphthenecarboxylic acid, 1-chloro-2-butanol and 4-methoxycyclohexanone. If two or more identical groups that qualify as principal group are present in the molecule, these are named with prefixes di-, tri-, etc., as in 2,3-hexanediol, 4-methoxy-1,3-cyclohexanediol, or 1,3,5-pentanetricarboxylic acid.

Table 4.1 GROUPS CITED ONLY AS PREFIXES IN SUBSTITUTIVE NOMENCLATURE

Group	Prefix	Group	Prefix
$-Br$	bromo	$=N_2$	diazo
$-Cl$	chloro	$-N_3$	azido
$-ClO$	chlorosyl	$-NO$	nitroso
$-ClO_2$	chloryl	$-NO_2$	nitro
$-ClO_3$	perchloryl	$=N(O)OH$	*aci*-nitro
$-F$	fluoro	$-OR$	R-oxy
$-I$	iodo	$-SR$	R-thio (similarly R-seleno
$-IO$	iodosyl		and R-telluro)
$-IO_2$	iodyl (replacing iodoxy)	Hydrocarbon or heterocyclic	
$-I(OH)_2$	dihydroxyiodo	substituents	
$-IX_2$	X may be halogen or a radical, and the prefix names are dihalogenoiodo, etc., or, for radicals, patterned on di-acetoxyiodo		

Table 4.2 SOME GENERAL CLASSES OF COMPOUND IN THE ORDER IN WHICH THE RELEVANT GROUPS HAVE DECREASING PRIORITY FOR CITATION AS PRINCIPAL GROUP

1. 'Onium and similar cations
2. Acids: in the order COOH, C(=O)OOH, then successively their S and Se derivatives, followed by sulfonic, sulfinic acids, etc.
3. Derivatives of acids: in the order anhydrides, esters, acyl halides, amides, hydrazides, imides, amidines, etc.
4. Nitriles (cyanides), then isocyanides
5. Aldehydes, then successively their S and Se analogs; then their derivatives
6. Ketones, then their analogs and derivatives, in the same order as for aldehydes
7. Alcohols, then phenols; then S and Se analogs of alcohols; then esters of alcohols with inorganic acids*; then similar derivatives of phenols in the same order
8. Hydroperoxides
9. Amines; then imines, hydrazines, etc.
10. Ethers; then successively their S and Se analogs
11. Peroxides

* Except esters of hydrogen halides (see *Table 4.1*)

Frequently a compound contains more than one type of group listed in *Table 4.2*, and since it is a rule that only one type can be named as suffix an order of priority is required; that order is the order in which the classes are listed in *Table 4.2*. The order appears arbitrary but it is based on a study made many years ago by *Chemical Abstracts* of the majority usage by chemists at that time when no official order existed.

Of course, any other substituents from this list that are present in the compound and not selected as principal group must be named as prefixes; so *Table 4.3* lists the names of the most important of these

Table 4.3 SUFFIXES AND PREFIXES FOR SOME IMPORTANT GROUPS IN SUBSTITUTIVE NOMENCLATURE

Class	Formula*	Prefix	Suffix
Cations		-onio	-onium
		-onia	—
Carboxylic acid	–COOH	carboxy	-carboxylic acid
	–(C)OOH	—	-oic acid
Sulfonic acid	–SO₃H	sulfo	-sulfonic acid
Salts	–COOM	—	metal . . .carboxylate
	–(C)OOM	—	metal . . .oate
Esters	–COOR	R-oxycarbonyl	R . . .carboxylate
	–(C)OOR	—	R . . .oate
Acid halides	–CO–Halogen	haloformyl	-carbonyl halide
	–(C)O–Halogen	—	-oyl halide
Amides	–CO–NH₂	carbamoyl	-carboxamide
	–(C)O–NH₂	—	-amide
Amidines	–C(=NH)–NH₂	amidino	-carboxamidine
	–(C)(=NH)–NH₂	—	-amidine
Nitriles	–C≡N	cyano	-carbonitrile
	–(C)≡N	—	-nitrile
Aldehydes	–CHO	formyl	-carbaldehyde
	–(C)HO	oxo	-al
Ketones	＞(C)=O	oxo	-one
Alcohols	–OH	hydroxy	-ol
Phenols	–OH	hydroxy	-ol
Thiols	–SH	mercapto	-thiol
Hydroperoxides	–O–OH	hydroperoxy	—
Amines	–NH₂	amino	-amine
Imines	=NH	imino	-imine
Ethers	–OR	R-oxy	—
Sulfides	–SR	R-thio	—
Peroxides	–O–OR	R-dioxy	—

* Carbon atoms enclosed in parentheses are included in the name of the parent compound and not in the suffix or prefix.

groups as both suffix and prefix. Simple examples are 2-aminoethanol, *o*-aminophenol, 2-oxocyclopentanecarboxylic acid and 3-ethoxy-4-fluorobenzamide.

Principal groups in radicofunctional nomenclature

In radicofunctional nomenclature we are dealing with names that consist of two or three words, the last word stating the function and the other(s) specifying the rest of the molecule in radical form. *Table 4.4* lists the commonest of these classes, again in priority order that governs the choice if more than one of these classes of group is present in one molecule. Examples are ethyl alcohol, ethyl chloride, phenyl azide and methyl cyanide.

Table 4.4   SOME FUNCTIONAL CLASS NAMES USED IN RADICOFUNCTIONAL NOMENCLATURE, IN ORDER OF DECREASING PRIORITY FOR CHOICE AS SUCH

Group	Functional class name
X in acid derivatives RCO–X, RSO$_2$–X, etc.	Name of X; in the order fluoride, chloride, bromide, iodide; cyanide, azide, etc.; then their S, followed by their Se analogs
–CN, –NC	Cyanide, isocyanide
>CO	Ketone, then S, then Se analogs
–OH	Alcohol; followed by S and then Se analogs
–O–OH	Hydroperoxide
>O	Ether or oxide
>S, >SO, >SO$_2$	Sulfide, sulfoxide, sulfone
>Se, >SeO, >SeO$_2$	Selenide, selenoxide, selenone
–F, –Cl, –Br, –I	Fluoride, chloride, bromide, iodide
–N$_3$	Azide

Any substituents not denoted by the functional class name are specified by prefixes, using the same prefix names as for substitutive nomenclature (*Table 4.3*) as, for example, in *p*-bromobenzyl cyanide.

When such a function is represented by a divalent formula (e.g. –O–, >CO), different groups attached at these two bonds are stated as separate words in alphabetical order; or if the two groups are identical their name is preceded by di-. Examples are benzyl phenyl ether, diethyl ether, benzyl 1-naphthyl sulfide and diethyl sulfoxide.

The reader may have noticed that -carboxylic acid or -oic acid* is actually a radicofunctional name, since acid denotes the function and the preceding word specifies the radical in adjectival form; acids appear in both *Tables 4.3* and *4.4*; the reason why they have to be included in substitutive nomenclature is the extremely high priority for citation as suffix that is assigned to them. Perhaps this is also the place to note that it is not the practice to call organic acids by names such as hydrogen acetate which would parallel the inorganic practice of hydrogen chloride, hydrogen tetrachloroaurate, etc.

It is obvious that none of the *Tables 4.2* to *4.4* approaches completeness, for there are vague phrases such as 'their derivatives' and 'etc.'; the variety in organic chemistry makes that inevitable for arbitrary priorities such as these; completeness would require an alphabetical order or some form of computer logic. Yet it is surprising how very seldom the need for finer distinctions arises in practice.

ALPHABETICAL ORDER

Reference has already been made several times above to alphabetical order in names and it is time now to define it precisely. The rules are quite simple.

Prefixes are arranged in alphabetical order, any multiplying parts of simple prefixes being neglected for this purpose. The atoms and groups are alphabetized first and the multiplying prefixes are then inserted, as in:

> *o*-**b**romochlorobenzene
> 4-**e**thyl-3-**m**ethyldecane
> 1,1,1-trichloro-3,3-di**m**ethylpentane

A complex radical forms one suffix; it is therefore alphabetized under its first letter, as in:

> 1-(**d**imethylamino)-3-ethyl-4-(**m**ethylamino)-2-naphthoic acid
> 4-**c**hloro-1,5-bis(**d**imethylamino)-3-ethyl-2-naphthoic acid

For otherwise identical prefixes the one with the lowest locant at the first cited point of difference is given first, as in:

> 1-(2-ethyl-**3**-methylpentyl)-8-(2-ethyl-**4**-methylpentyl)naphthalene

* The difference between these names is explained in Chapter 6.

When two words fill identical functions in a three-word name they are alphabetized, as in:

ethyl **m**ethyl ketone
butyl **e**thyl ether

Italics are neglected in alphabetizing, as in:

3-(*trans*-2-**b**utenyl)-2-**e**thylphenol

NUMBERING

The principles of numbering (enumeration) used for compounds and for radicals are the same, with one exception, namely: for compounds the lowest available numbers are assigned to the principal (functional) group or groups; for radicals they are assigned to the 'free' valence or valences, and the principal groups are then treated merely as ordinary prefixes.

'Lowest numbers' has a specific meaning in nomenclature. When two or more sets of numbers are compared in ascending order of numbers, that set is 'lowest' which contains the lowest individual number on the first occasion of difference. Thus 1,1,7,8 is lower than 1,2,3,4 [see also examples (10) and (12)].

The international rules on the subject state that for aliphatic compounds lowest numbers are assigned successively, so far as applicable and until a final decision is reached, to (i) principal groups, (ii) unsaturation (i.e. double and triple bonds together), (iii) double bonds, (iv) triple bonds, (v) atoms or groups designated by prefixes, (vi) prefixes in order of citation (in Great Britain and USA, alphabetical).

For ring systems there is a prescribed numbering (outlined in Chapter 5) which often leaves little or no room for choice according to the substituents present. However, so far as choice remains, lowest numbers are given to (i) 'indicated' hydrogen (see pp. 73, 112), (ii) principal groups, (iii) multiple bonds in compounds whose names indicate partial hydrogenation (cycloalkenes, pyrazolines, and the like), (iv) substituents named as prefixes (including hydro), (v) prefixes in order of citation (which is alphabetical in USA and Great Britain).

The simplest case is when the numbering of the parent compound is entirely fixed in advance, as occurs with many cyclic compounds (see Chapter 5). Substituents in, say, indole, can have only one number each; compound (1) must be 5-chloro-1-methyl-3-indole-carboxylic acid.

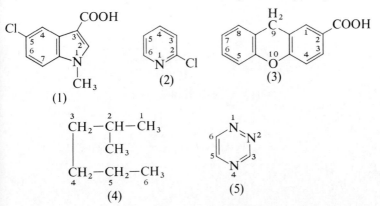

(1) (2) (3) (4) (5)

Next there is the simple case of one group (which need not be a principal group) in a compound such as (2), (3), (4) or (5) which can be numbered in either of two ways: the lower numbers for substituents are as shown and are to be used.

If the ring leaves the numbering completely free, e.g. (6), the principal group begins the numbering. If there is more than one

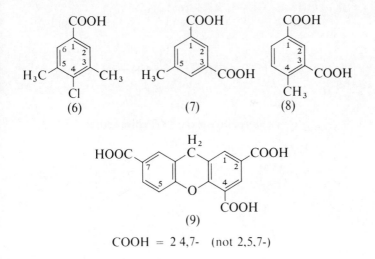

(6) (7) (8)

(9)

COOH = 2 4,7- (not 2,5,7-)

principal group, they may be required together to decide the numbering, and may do so either wholly as in (7), or partly as in (8) where the methyl group decides which carboxyl group shall have number 1 and which number 3. Our definition of smallest numbers leads to decisions as shown in (9)–(13).

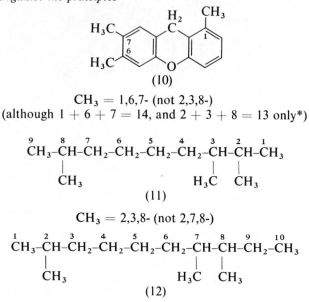

(10)

CH₃ = 1,6,7- (not 2,3,8-)
(although 1 + 6 + 7 = 14, and 2 + 3 + 8 = 13 only*)

$$\overset{9}{C}H_3-\overset{8}{C}H-\overset{7}{C}H_2-\overset{6}{C}H_2-\overset{5}{C}H_2-\overset{4}{C}H_2-\overset{3}{C}H-\overset{2}{C}H-\overset{1}{C}H_3$$
$$\qquad\quad|\qquad\qquad\qquad\qquad\qquad\quad|\quad\ |$$
$$\qquad\quad CH_3 \qquad\qquad\qquad\qquad\qquad H_3C\ \ CH_3$$

(11)

CH₃ = 2,3,8- (not 2,7,8-)

$$\overset{1}{C}H_3-\overset{2}{C}H-\overset{3}{C}H_2-\overset{4}{C}H_2-\overset{5}{C}H_2-\overset{6}{C}H_2-\overset{7}{C}H-\overset{8}{C}H-\overset{9}{C}H_2-\overset{10}{C}H_3$$
$$\qquad\quad|\qquad\qquad\qquad\qquad\qquad\quad|\quad\ |$$
$$\qquad\quad CH_3 \qquad\qquad\qquad\qquad\qquad H_3C\ \ CH_3$$

(12)

CH₃ = 2,7,8- (not 3,4,9-)
(although 2 + 7 + 8 = 17, and 3 + 4 + 9 = 16 only*)

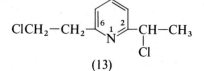

(13)

1-chloroethyl precedes 2-chloroethyl

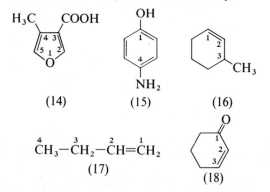

(14)   (15)   (16)

$$\overset{4}{C}H_3-\overset{3}{C}H_2-\overset{2}{C}H=\overset{1}{C}H_2$$

(17)

(18)

* There has been a belief in some quarters that the total of numbers must be he smallest possible: that is erroneous.

Lastly we can run quickly through the international sequence. The principal group decides in cases such as (14) and (15). If there is no principal group, a double bond cited as -ene decides, as in (16) and (17); also it can make the second choice if the principal group is not decisive, as in (18); but note, however, that unsaturation not cited as -ene is not relevant here: hydro- prefixes are treated in the same way as any other prefix (see below). Certain heterocycles have modified names when partly hydrogenated (see pp. 84–86), and then too the unsaturation decides the numbering. If there is no principal group or -ene double bond, triple bonds are considered, as in (19) and (20). Next, prefixes are considered, first all together, independently of the kind and including hydro-. Examples are (21;

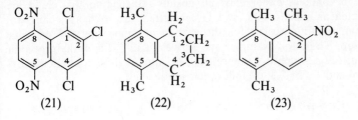

$$\overset{5}{CH_3}-\overset{4}{CH}(CH_3)-\overset{3}{C}\equiv\overset{2}{C}-\overset{1}{CH_3}$$

(19)

(20)

1,2,4,5,8-, not 1,4,5,6,8-), (22; 1,2,3,4-tetrahydro-) and (23; 1,2,5,8-, not 1,4,5,6-). All other things being equal, the lower number goes to the prefix cited first in the name, as in (24; bromo before chloro),

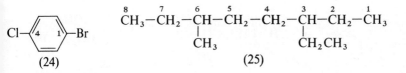

(21)   (22)   (23)

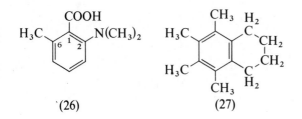

(24)

$$\overset{8}{CH_3}-\overset{7}{CH_2}-\overset{6}{CH}-\overset{5}{CH_2}-\overset{4}{CH_2}-\overset{3}{CH}-\overset{2}{CH_2}-\overset{1}{CH_3}$$

(25)

(26)

(27)

(25; ethyl before methyl), (26; dimethylamino before methyl) and (27; tetrahydro before tetramethyl).

That completes the general rules. Problems, largely theoretical, arising from severe ramification of aliphatic chains are considered in the IUPAC rules[1a] but need not concern us here.

CONJUNCTIVE NOMENCLATURE

This type of nomenclature (earlier called 'additive nomenclature' in *Chemical Abstracts*), whose considerable value is too slowly receiving general recognition, can be applied where a ring system (other than a benzene ring) is attached through carbon to a carbon atom of an aliphatic chain that carries a principal group. The name of the ring system is followed directly by the name of the aliphatic chain and principal group, as in 2-naphthaleneethanol or 2-naphthaleneacetic acid. Two hydrogen atoms are assumed to be lost in this process, and not one as in substitutive nomenclature. Locants used for the ring system are the usual numerals—the '2' in the examples above refers to the 2-position of the naphthalene nucleus and not that of the ethanol or acetic acid. For this system of nomenclature the aliphatic chain named runs from the ring system to the functional group—and not beyond, at either end; all groups attached to this chain are treated as substituents, and Greek letters are used as locants for them, starting with α for the carbon atom bearing the principal group. Thus, for example, one arrives at (28) 2-naphthylpropionic acid, (29) 2-naphthylethanol, (30) α,γ-dimethyl-2-naphthalenepropanol and (31) 1,2,3,-cyclohexanetriacetic acid.

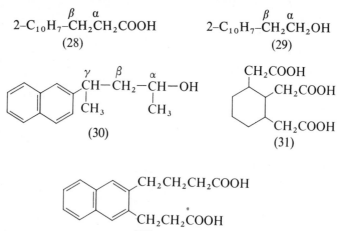

$$\overset{\beta}{\phantom{2-C_{10}H_7-}}\overset{\alpha}{}$$
2–$C_{10}H_7$–CH_2CH_2COOH
(28)

$$\overset{\beta}{\phantom{2-C_{10}H_7-}}\overset{\alpha}{}$$
2–$C_{10}H_7$–CH_2CH_2OH
(29)

$$\overset{\gamma}{CH}-\overset{\beta}{CH_2}-\overset{\alpha}{CH}-OH$$
$$\overset{|}{CH_3}\qquad\overset{|}{CH_3}$$
(30)

CH_2COOH
CH_2COOH
CH_2COOH
(31)

$CH_2CH_2CH_2COOH$
CH_2CH_2COOH
(32)

If the side chains are not identical, one (the simpler) is named as prefix. For example, (32) would be called 3-(2-carboxyethyl)-2-naphthalenepropionic acid.

To avoid numbering difficulties, conjunctive nomenclature is not used with unsaturated side chains or when a principal group occurs more than once in a single side chain (except for malonic and succinic acid).

Further detail on rare cases will be found in the 1970 IUPAC rules[1b].

This method is recommended for greater use by European chemists.

A CLOSER APPROACH TO A NAME

The ways in which a name should be built were given in outline on pp. 49–51. Now that principal groups and numbering have been discussed, the outline can be filled in.

The starting point remains the principal group, as it was for numbering. This identity is very important, for it ensures that the name and the numbering shall run on parallel lines. (If there is no principal group in the compound, the name is approached simply as described in the next chapter, and prefixes are added later.)

For aliphatic compounds the complete chain is the parent to which the suffix denoting the principal group is added, as in 2-hexanol. If there is a choice of chains, priority is given to the most unsaturated,

$$CH_3CH{=}CCH_2OH \qquad\qquad CH_3CH_2CHCH_2OH$$
$$\mid \qquad\qquad\qquad\qquad\qquad \mid$$
$$CH_2CH_2CH_3 \qquad\qquad\qquad CH_2CH_2CH_3$$

(33)   (34)

as in 2-propyl-2-buten-1-ol (33), even though this may not be the longest. (Double bonds have preference over triple bonds.) If there is a choice between chains of equal degree of hydrogenation, then, of course, the longest chain is chosen, as in 2-ethyl-1-pentanol (34), even though as a result the related compounds (33) and (34) are then named from different parent hydrocarbons.

It will be remembered that the principal group is the group highest in *Table 4.3* (p. 53), and that all others, including hydro*, are given as prefixes, all in alphabetical order. This division into suffix and prefixes can cause chemically unwelcome changes in name or numbering, or both: for instance, compound (35) is 3-hexanol, but (36) is 4-hydroxy-3-hexanone. The same sort of change, can, however,

---

* This does not apply in German.

result from shift of a double bond: (37) is 2-ethyl-1-hexene, but (38) is 5-methyl-2-heptene. The fact is that names are founded on structures, not on chemical relations between two substances; the reason, of course, is that it is often easy to find relationship to more than one other substance so that choice of a name would then become ambiguous.

$$\overset{3}{C}H_3CH_2CH_2\overset{2}{C}HCH_2\overset{1}{C}H_3$$
$$\underset{OH}{|}$$
(35)

$$\overset{1}{C}H_3CH_2\overset{2}{C}-\overset{3}{C}H\overset{4}{C}H_2CH_3$$
$$\underset{O\ OH}{\|\ |}$$
(36)

$$\overset{3}{C}H_3CH_2CH_2\overset{2}{C}H_2\overset{1}{C}{=}CH_2$$
$$\underset{CH_2CH_3}{|}$$
(37)

$$\overset{1}{C}H_3\overset{2}{C}H{=}\overset{3}{C}HCH_2\overset{4}{C}H\overset{5}{C}HCH_3$$
$$\underset{CH_2CH_3}{|}$$
(38)

There is no difficulty in citing multiple groups of one kind attached to a single chain or ring system; 3,4-hexanediol and 1,2,3,4-benzenetetracarboxylic acid admit no argument. With branched chains further rules come into play. Substance (39) must be named so that both functional groups can become suffixes to the main chain: the name is 2-butyl-1,4-pentanediol and not 2-(2-hydroxypropyl)-1-hexanol, even though the latter involves a hexane chain and the correct name involves only a pentane chain.

$$CH_3CH_2CH_2CH_2CHCH_2OH$$
$$\underset{CH_2CH(OH)CH_3}{|}$$
(39)

$$HOCH_2CH_2CHCH_2CH_2OH$$
$$\underset{CH_2CH_2OH}{|}$$
(40)

The principle of the preceding paragraph is often spoken of as 'treating like things alike'. In general that is indeed a good thing to do; but it cannot be made into a guiding light for nomenclature, for there is often argument about degrees of 'likeness', and in any case the principle itself must often be discarded. Consider, for example, the simple triol (40). Here the name must be based on a pentane chain containing two hydroxyl groups, leaving the third hydroxyl group in a side chain: the correct name is 3-(2-hydroxyethyl)-1,5-pentanediol. Now there could be a system of nomenclature by which this substance would be called 2-ethylpentane-1,5,2'-triol, but in fact such a system is not used and this name is incorrect.

Wholly symmetrical aliphatic compounds such as (40) are, however, rather rare. More commonly, branched chains offer more alternatives. Consider a choice between various chains containing the same number of identical groups. As when there was only one principal group, the additional rule favoring unsaturation requires that compound (41) be named 4-(4-hydroxybutyl)-2-octene-1,8-diol, even though selection of the two saturated branches would give a

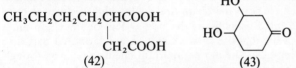

(41)

nonane derivative. When there is no difference in degree of saturation the longer chain would be chosen, and the saturated analog of the octene (41) would be 5-(3-hydroxypropyl)-1,9-nonanediol.

There are occasions when further choices are needed for highly branched chains, but they become too rare to need discussion here. However, before passing on, let us note that the acid (42) can be called butylsuccinic acid or 2-butyl-1,4-butanedioic acid; whether the trivial succinic or the systematic butanedioic is used, the same principles are applied, leading here to the C_4 chain, not the C_6 chain, as the basis of the name.

CH₃CH₂CH₂CH₂CHCOOH
　　　　　　|
　　　　　CH₂COOH

(42)

(43)

The rule about seniority for principal groups, in the order on p. 53, is independent of the relative numbers of the various groups. The compound (43) is 3,4-dihydroxycyclohexanone, and not 4-oxo-1,2-cyclohexanediol: this is in line with the numbering rules which give the lowest available number to the principal group even though the sequence 1,2,4 is 'lower' than 1,3,4. There are many and varied types of problem in which this consideration arises: but the answer is always the same, so further elaboration is unnecessary.

The procedures discussed above are applicable equally when the principal groups are attached directly to cyclic systems. With the more complex ring systems, derivation of the full name is, in fact, usually simpler (once the cyclic parent has been named) because the numbering is at least partly fixed. With simpler ring systems, and particularly the aromatic ones, the systematic procedures become more troublesome, partly from older custom and partly because there is here a very large number of trivial names which should reasonably be carried over to the derivatives.

Strict application of the rule that principal groups shall be cited as suffix is relatively recent, particularly for cyclic compounds, and there are many situations where the resulting names may appear unfamiliar to, or even shock, some older chemists; 2,3-dihydroxy-naphthalene and 2-aminophenanthrene, for instance, may appear more usual to some than the correct 2,3-naphthalenediol and 2-phenanthrenamine. And there is a further stumbling block with derivatives of *N*-heterocyclic systems; the -ine of, say, pyridine may be considered a suffix prescribing the principal group, but if a 'senior' group in the sense of *Table 4.3*, say –OH or –COOH, is present, then that should be stated as suffix, a procedure that gives in these cases 2-pyridinol and 4-pyridinecarboxylic acid. These are IUPAC and *Chemical Abstracts* names, so it seems wrong to regard the -ine of pyridine as a suffix, and if that is so, then names such 2-pyridinamine become perfectly satisfactory, and that is indeed the case; that is the modern name whose use is advisable.

With compounds containing both chains and rings the position is simple when the functional groups are directly attached either all to the chain or all to the ring system; these groups decide the parent name, e.g. 3-phenyl-1-propanol but *m*-propylphenol. And conjunctive nomenclature should never be forgotten if the cyclic component is at all complex. Difficult cases can arise when there are principal groups attached to both the chain and to the ring system, as they cannot all be treated as suffixes: the ring system or the chain must be chosen as parent, and the choice will usually fall on that part that is the more complex or the more important chemically, even though these criteria are both matters of opinion. Few would quarrel with names such as 8-(*p*-hydroxyphenyl)-2,4,6-octatrien-1-ol or 1-(*p*-hydroxyphenyl)glucose; but it will depend on the chemistry under discussion whether a name such as 3-(*p*-hydroxyphenyl)-1-propanol or *p*-(3-hydroxypropyl)phenol is chosen.

Care is similarly required when principal groups are present in more than one chain or in more than one ring system forming part of a single compound. IUPAC has laid down criteria as to the 'seniority' between various chains and between various rings; in general the more complex is the senior, but the rules[1c] must be studied for detail.

Finally there is the nomenclature that takes advantage of the fact that organic synthesis often leads to symmetrical compounds of type X–Y–X where the same principal group is present in both X groups but not in Y. When X and Y are not both aliphatic or both cyclic, or when Y is a hetero group, simple and readily intelligible names can be devised by a special elaboration of the standard rules, as mentioned briefly on p. 50. The examples given were:

$$p\text{-HOOCCH}_2\text{-C}_6\text{H}_4\text{-CH}_2\text{COOH} \qquad (44)$$

and $\qquad \text{H}_2\text{NCH}_2\text{CH}_2\text{-O-CH}_2\text{CH}_2\text{NH}_2 \qquad (45)$

In (44) the two acetic acid residues are cited as acetic acid and 'doubled' by the prefix di-; the C_6H_4 group called phenylene is then placed in front, giving *p*-phenylenediacetic acid, which describes the structure as clearly as phenylacetic acid describes $\text{C}_6\text{H}_5\text{CH}_2\text{COOH}$. Similarly, compound (45) is called 2,2'-oxydi(ethylamine), where the parentheses differentiate the two ethylamine residues from diethylamine $(\text{C}_2\text{H}_5)_2\text{NH}$. This procedure is quite widely applicable, for a variety of groups can function as Y; for example, $>\text{CO}$ carbonyl, $-\text{CH}_2-$ methylene, $-\text{CH}_2\text{CH}_2-$ ethylene, $-\text{CH}_2\text{CH}_2\text{CH}_2-$ trimethylene (etc.), $-\text{NHCONH}-$ ureylene, $-\text{S}-$ thio, $>\text{CS}$ thiocarbonyl, $>\text{NH}$ imino, $-\text{N}{=}\text{N}-$ azo. Further, the principle can be extended to tervalent radials, e.g. $\text{N(CH}_2\text{COOH)}_3$ is nitrilotriacetic acid; and, finally, the Y group may itself be complex and symmetrical as in (46), which is named 4,4'-methylenedioxydibenzoic acid. A warning

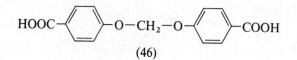

$$\text{HOOC}\!-\!\!\!\left\langle\ \right\rangle\!\!\!-\!\text{O}-\text{CH}_2-\text{O}\!-\!\!\!\left\langle\ \right\rangle\!\!\!-\!\text{COOH}$$

(46)

is, however, needed: trouble will arise if there is not complete symmetry of the two X groups or within Y, and this nomenclature should not be attempted in such cases.

It will be clear, then, how the building of a name is not always simply a following of rules. Though in a truly surprisingly high proportion of cases the rules and common trivial names suffice, and though, let it be emphasized again, rules should not be broken without good cause, yet chemistry must not be forgotten. The versatility of organic nomenclature in fact arises from the reasonable wish to express chemical relations in a name; and in this too the difficulties originate, as when multiple chemical relations lead to conflicting types of name or when a specialized usage is extended beyond its appropriate field; then it is wise to revert to the systematic types, which thus remain the backbone of nomenclature. It is essential to appreciate both the uses and the limitations of systematics before indulging in excursions from it.

REFERENCE

1. *IUPAC Nomenclature of Organic Chemistry, Definitive Rules for: Section A. Hydrocarbons; Section B. Fundamental Heterocyclic Systems; Section C. Characteristic Groups Containing Carbon, Hydrogen, Oxygen, Nitrogen, Halogen, Sulfur, Selenium and/or Tellurium.* 1969. A,B 3rd edn.; C 2nd edn.; Butterworths, London (1971). (a) pp. 8–11, 13–15; (b) pp. 119–123; (c) pp. 97–105

5 Organic: Hydrocarbons and Heterocycles

ALKANES

Standard nomenclature of alkanes is well known, but a few special points are worth noting.

A prefix iso, to denote terminal $(CH_3)_2CH-$, is restricted to C_4-C_6 alkanes and to C_3-C_6 radicals; *tert*- is restricted to C_4 and C_5 alkanes and their radicals; *sec*- is restricted to *sec*-butyl, and neo to neopentane $C(CH_3)_4$ and neopentyl $(CH_3)_3CCH_2-$; *tert*- and *sec*- are italicized and followed by a hyphen, iso and neo are not; *n*-, if used to specify normal (which should not be done except for contrast), should be italicized and followed by a hyphen as it is not to be used in indexing.

Iso, *sec*- and *tert*- names should not be used for substituted radicals because difficulties with numbering then arise.

In highly branched alkanes choice of the 'main chain', i.e. that chain which is regarded as substituted by the other chains, can present considerable difficulties, at least in hypothetical compounds. Complex rules have been provided by IUPAC to cover such cases[1a].

Unsaturation, as is well known, is indicated by replacing the ending -ane by -ene (for C=C) or -yne (for C≡C), and both may occur in a name (-ene before -yne with elision of 'e' before a vowel), e.g. 1-buten-3-yne or 2-hexadecene-5,7-diyn-4-ol.

Compounds containing $-C=C-C=C-$ groups (or more extended forms) are said to have conjugated double bonds; $-C=C-C≡C-$ groups are also said to have conjugated multiple bonds. Compounds containing $-C=C=C-$ or similar but more extended systems are said to have cumulative unsaturation and are generically called cumulenes; they are numbered normally, e.g. 1,2-butadiene $CH_3C=C=CH_2$.

CYCLOALKANES

The nomenclature of cycloalkanes and their unsaturated derivatives follows closely that of unbranched alkanes; all substituents attached to the ring are named as such. Their numbering follows the principles discussed in Chapter 4, including the rule that points of attachment of cycloalkenyl groups named as such, take preference over unsaturation for lower number, e.g. 2-cyclohexenyl (and not 1-cyclohexen-3-yl) is the correct name for the radical (1).

(1)

BENZENE DERIVATIVES

The IUPAC rules follow general custom. A few points that still entice occasional mistakes may be mentioned.

It is no longer customary to assign lowest numbers to methyl substituents of simple benzene derivatives; the standard rules of Chapter 4 apply, so that, for example, 2-chloro-*p*-toluidine describes the compound having the chlorine atom next to the amino group.

The following names and particularly the Greek letters identifying the side-chain atoms are official:

$\overset{\alpha}{C_6H_5-CH_3}$	toluene	$\overset{\gamma}{C_6H_5}\overset{\beta}{CH}=\overset{\alpha}{CH}-\overset{}{CH_2}-$	cinnamyl
$CH_3C_6H_4-$	tolyl	$C_6H_5\overset{\alpha}{CH}=\overset{\beta}{CH_2}$	styrene
$\overset{\alpha}{C_6H_5CH_2-}$	benzyl	$C_6H_5\overset{\beta}{CH}=\overset{\alpha}{CH}-$	styryl
$(\overset{\alpha}{C_6H_5})_2CH-$	benzhydryl		
$(\overset{\alpha}{C_6H_5})_3C-$	trityl		

The xylyl radical is $(CH_3)_2C_6H_3-$; a radical $CH_3C_6H_4CH_2-$ is methylbenzyl.

There is no such radical as cresyl (from cresol): it is tolyl (*pace* tricresyl phosphate).

BI-COMPOUNDS

Bi- has long been American usage to denote direct union of two rings, cf. (2) and (3), which in Europe were earlier called diphenyl and 2,2′-dinaphthyl. The IUPAC rules early adopted bi- (in place of di-), and this prefix is now generally accepted; it is exclusive and unique for 'doubled groups' (except for bicyclo, see below). The IUPAC rules, however, permit doubling of either the group, giving binaphthyl, or of the molecule, giving binaphthalene; in the latter case the use of bi- carries the implication that two hydrogen atoms are lost as in conjunctive nomenclature (p. 60). Doubling of the molecule is now almost universally accepted; but so is biphenyl, as an exception of very long standing (note that the radical C_6H_5–C_6H_4– is biphenylyl, the second 'yl' being the termination for a radical).

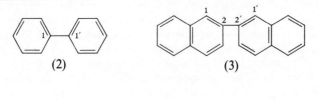

(2)   (3)

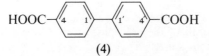

(4)

Numbering is as shown in formulas (2) and (3), the points of union having the lowest available numbers.

The 'bi' nomenclature is reserved for unsubstituted compounds in the sense that only parent compounds may be 'doubled'; for example, compound (4) is 4,4′-biphenyldicarboxylic acid, not *p,p′*-bibenzoic acid (as it was earlier in America).

Ter, quater, etc., are used for larger ring assemblies, always with the name of the molecule (e.g. ternaphthalene, quaterpyridine), but again with terphenyl, quaterphenyl, etc., as exceptions. For assemblies of the same unit, unprimed numbers are assigned to one of the terminal units, the nearest neighbor unit has singly primed and the next doubly primed numbers, and so on; as usual, points of union have the lowest numbers possible, these numbers being considered an attribute of the parent structure and not subject to change by presence of a suffix—e.g. the COOH of (4) are numbered 4 and not 1.

POLYCYCLIC HYDROCARBONS

The nomenclature of polycyclic hydrocarbons, which is the basis also for that of heterocycles, is complex because of the multitude of compounds and different situations that can be met. Chemists should become familiar with the basic principles, which are all that can be described here. There is a considerable list of prescribed names, a method (the fusion method) of building names for more complicated systems derived from members of that list, a second method of building polycyclic names (the bicyclo system) independent of the fusion principle and the prescribed list, plus a few special methods for specific types of compound. We must deal with them in that order, but, before doing that, it will be best to note the standard method of orienting and numbering polycyclic compounds.

(i) Whenever possible, rings are drawn with two sides vertical (a three-membered ring with one vertical). (ii) As many rings as possible are then drawn in a horizontal line, irrespective of their size. (iii) As much as possible of the remainder of the formula is arranged in the top right quadrant, and as little as possible in the bottom left quadrant (the middle of the first row being the center of the circle). (iv) *Then*, numbering proceeds clockwise round the periphery, starting in the right-hand ring of the top row at the first carbon atom not engaged in ring fusion. Atoms engaged in ring fusion (i.e. at the 'valley' positions) receive roman letters a, b, etc., after the numerals of the preceding atom. (Subsidiary rules legislate for less common points.)

Examples are chrysene (5) and rubicene (6).

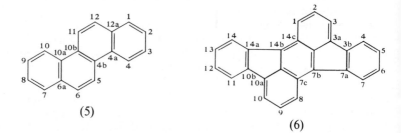

(5)

(6)

'Prescribed' polycycle names

Table 5.1 reproduces the IUPAC list of 35 carbopolycycles whose names can be used for the fusion method of further nomenclature. A few points concerning this list are important. Except for the long-established anthracene and phenanthrene, the numbering is that derived by the standard procedure just given. All the compounds are in the oxidation state where they contain the maximum number of non-cumulative double bonds, and the ending -ene of the names of all ring systems should denote this (and not a single double bond as for aliphatic compounds). Some cyclic skeletons require a CH_2 component in the ring; these components are to be indicated by

Pentalene

Indene

Naphthalene

Azulene

Heptalene

Biphenylene

as-Indacene

s-Indacene

Acenaphthylene

Fluorene

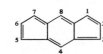
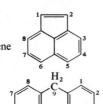

11. Phenalene

12. Phenanthrene

13. Anthracene

14. Fluoranthene

15. Acephenanthrylene

16. Aceanthrylene

17. Triphenylene

18. Pyrene

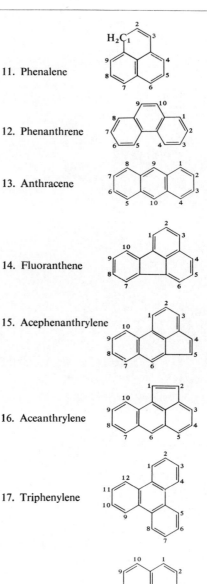

(contd. on p. 72)

72

Table 5.1 (contd.)

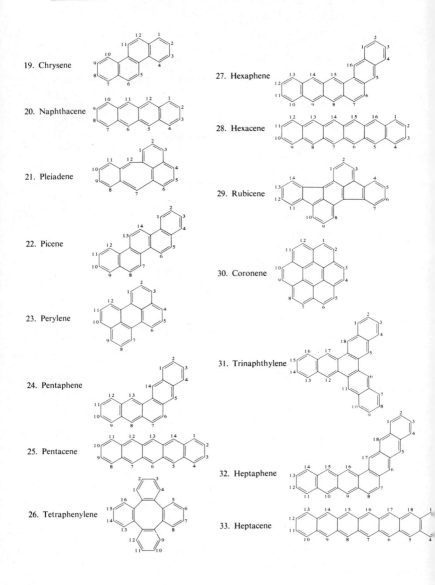

19. Chrysene

20. Naphthacene

21. Pleiadene

22. Picene

23. Perylene

24. Pentaphene

25. Pentacene

26. Tetraphenylene

27. Hexaphene

28. Hexacene

29. Rubicene

30. Coronene

31. Trinaphthylene

32. Heptaphene

33. Heptacene

Table 5.1 (contd.)

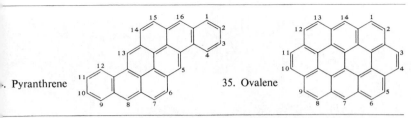

. Pyranthrene

35. Ovalene

italic *H* preceded by a numeral, except when in common compounds such as indene and fluorene its position can be assumed to be the normal one (see *Table 5.1*); if the CH_2 group were in a different position, say, as in compound (7), the name would be 3*H*-fluorene.

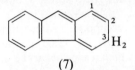

(7)

Such hydrogen is termed 'indicated hydrogen'; it is discussed in greater detail on p. 112.

Hydrogenation is normally denoted by hydro- prefixes (including perhydro- for complete hydrogenation), but the seven partly hydrogenated systems (8) to (14) have recognized specific names (not to be

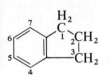

(8) Indan

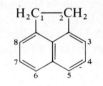

(9) Acenaphthene

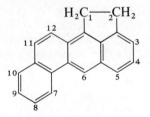

(10) Cholanthrene

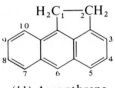

(11) Aceanthrene

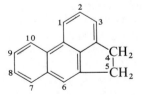

(12) Acephenanthrene

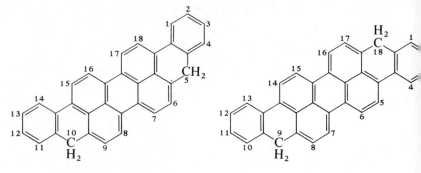

(13) Violanthrene   (14) Isoviolanthrene

used for ring fusion). Note the relation of indan to indene and of the
ace . . . ene to the ace . . . ylene names.

Table 5.1 includes six compounds whose names have been systema-
tized starting from the trivial name naphthacene (four benzene
rings fused in a row). Pentacene, hexacene and heptacene contain,
respectively, 5, 6 and 7 benzene rings fused in a row. Pentaphene,
hexaphene and heptaphene are dog-leg analogs.

The fusion (annelation) method for hydrocarbons

Aromatic polycyclic hydrocarbons not in the IUPAC list (*Table 5.1*)
are named by fusion, for which also the IUPAC rules[1b] give the
Chemical Abstracts (*Ring Index*) system. So many situations are to
be covered that again the rules need consultation for complex cases,
but the following is an outline of first principles.

(a) The largest possible unit with a trivial name in the list is
chosen as base component (in case of doubt, the component shown
last in the list), and its faces are lettered *a* (italic) for face 1-2, and
then *b*, *c*, etc., consecutively around the periphery, as in (15). The
complete name is then built up with an 'o' affix and square brackets
as shown; the lowest available numbers (independently of substituents) are used for the fusion positions; and the numerals in the
brackets must be in an order to correspond with 1,2 for face *a*, etc.
The formula is then re-orientated, if necessary, to accord with rules
(i)–(iii) on p. 70 and *re-numbered* in accord with rule (iv) as shown in
formula (15).

(b) Case (16) is one of three-point fusion.

(c) Designation within the square brackets is easier for benzo
derivatives such as (17) and (18) (p. 76).

(d) Abbreviated names are used for the prefixes benzo, naphtho,

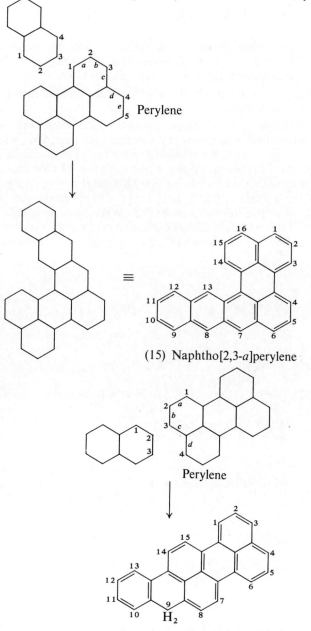

(15) Naphtho[2,3-*a*]perylene

Perylene

(16) 9*H*-Naphtho[1,2,3-*cd*]perylene

anthra (N.B. -a), phenanthro, acenaphtho and perylo, but not for others.

(e) The final name applies to the compound with the maximum number of non-cumulative double bonds; thus in indeno[1,2-a]indene (19) the characteristic CH_2 of indene has disappeared.

(f) Reduction products are indicated by hydro prefixes (perhydro for complete reduction).

(g) Cycloalkene rings, when present as terminal components, are denoted by prefixes cyclopenta, cyclohepta, cycloocta, etc., which refer to the rings with the maximum number of non-cumulative double bonds. Thus we arrive at the names shown for (20) and (21) [note in passing that the steroid nucleus (20) has the classical steroid (and not the general IUPAC) numbering]. The six-membered ring is named benzo also in cases such as (22). When fused terminally to one of the systems in *Table 5.1*, the affixes cyclobuta and cyclopropa may be used for the unsaturated C_4 and C_3 rings, as for example in

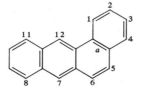

(17) Benz[a]anthracene   (18) 1H-Benz[de]anthracene

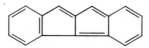

(19) Indeno[1,2-a]indene

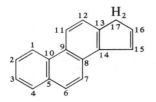 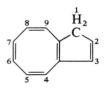

(20) Cyclopenta[a]phenanthrene   (21) 1H-Cyclopentacyclooctene

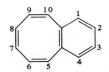

(22) Benzocyclooctene

(23) 4*H*-Cyclobutindene   (24) 4*H*-Bicyclo[7.2.0]undecapentaene

(23); actually such cases as (23) are rare; the IUPAC rule is so formulated that fusion principles are used only when two or more rings containing five or more ring members are fused together; all the compounds in *Table 5.1* are in accord with this, but in (24) there is only one ring with more than four ring members and such compounds must be named by the bicyclo method (see below).

(h) The 'o' indicating fusion is elided before a vowel, independently of brackets, as in benz[*a*]anthracene (17).

(i) The reason why square brackets are used for fusion locants is that the numerals do not correspond with those for the final numbering.

Bicyclo (etc.) nomenclature

This system was devised primarily for compounds containing bridged aliphatic rings, preferably saturated, where the resulting two or more rings had two, or often more, atoms in common. Typical simple examples are shown in formulas (25) and (26). The method

$$
\begin{array}{ccc}
{}^{7}CH_2-{}^{1}CH-{}^{2}CH_2 & & {}^{8}CH_2-{}^{1}CH-{}^{2}CH \\
| & {}^{8}CH_2\ {}^{3}CH_2 & {}^{7}CH_2 \qquad {}^{3}CH \\
{}^{6}CH_2-{}^{5}CH-{}^{4}CH_2 & & {}^{6}CH_2-{}^{5}CH-{}^{4}CH_2
\end{array}
$$

(25) Bicyclo[3.2.1]octane   (26) Bicyclo[3.3.0]oct-2-ene

has, however, also to be used for highly unsaturated systems such as (24) which do not contain two rings having five or more ring members. The names are produced as follows.

Count the number of carbon atoms separating one bridge end from the other by the various routes and place these numbers, separated by full stops, in descending order inside square brackets; in front of the brackets place bicyclo and after the brackets place the name of the alkane containing the total number of carbon atoms (now including the bridges), as shown under formulas (25) and (26). Numbering starts at one bridge end, goes to the other bridge end by the longest route, continues by the next longest route back to the first bridge end, and finally goes by the last (shortest) path to the second bridge end. Unsaturation is handled simply as in (26). The method is often extended to tricyclic and more complicated bridge structures, and seems to be the only method of naming cage structures in general. Further rules necessary for such extensions are given by IUPAC[1c].

Hydrocarbon-bridged (aromatic) systems

Hydrocarbon bridges can be named by a quite different procedure that is most useful for bridges across aromatic systems. The bridge is denoted by means of the corresponding hydrocarbon molecule name (not the radical name), with a terminal 'o', e.g. methano $-CH_2-$, ethano $-CH_2CH_2-$, benzeno $-C_6H_4-$, and is attached to the name of the ring system to be bridged. Compounds (27) and (28) provide two examples. Note two points even for such relatively simple systems: in both these examples, and in almost all other

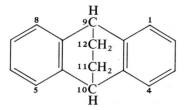

(27) 9,10-Dihydro-9,10-ethanoanthracene

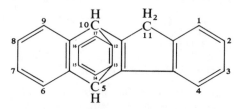

(28) 10,11-Dihydro-5,10-*o*-benzeno-5*H*-benzo[*b*]fluorene

examples met in practice, the bridged ring is reduced—one hydrogen atom at each bridge end—so the names must then contain hydro prefixes. Secondly, the bridge is numbered starting from the end nearest the previous highest number. For further examples which can provide tricky problems of nomenclature the IUPAC rules can be consulted[1d].

Spirans

Spirans are structures in which two ring systems share one common atom. As a result of IUPAC rules and general progress the various methods of naming them can now be simplified to the following two easy rules.

For two monocarbocycles in a spiran union the compound is named by placing spiro before the name of the single large ring containing the same *total* number of carbon atoms; immediately after 'spiro' are placed numerals denoting the numbers of atoms linked to the spiro atom in each ring; these numbers start with the lower and are separated by a full stop (period) and placed in square brackets. Numbering of the complete compound starts from the carbon atom of the smaller terminal ring next to the spiro atom. Thus (29) is named spiro[4.5]deca-1,6-diene, and a simple extension of the method gives the name dispiro[5.1.7.2]heptadecane for (30).

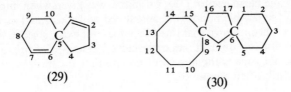

(29)   (30)

When one or more component is a fused polycyclic system the different methods exemplified by 1,1'-spirobiindene for (31) and spiro-[indene-1,1'-cyclopentane] for (32) are used.

(31)   (32)

Rings with side chains

A cyclic hydrocarbon with short side chains is normally treated as a substituted cyclic compound, e.g. hexamethylbenzene. A simple ring system with a long side chain is normally treated as a substituted aliphatic hydrocarbon, e.g. 1-phenyldecane. Complex ring systems are normally treated as parents, even if they have long side chains, e.g. 1-dodecylpyrene. Choice is free for intermediate cases (*Chemical Abstracts* considers pentyl short and hexyl long for benzene derivatives). Unsaturation in the side chain tends to favor the side chain as parent, e.g. 1-(2-naphthyl)hexa-1,3-diene. Chains substituted by two or more cyclic radicals are also most conveniently treated as parents, since, for example, 2-(2-naphthyl)-4-phenylhexane is 'simpler' than 2-(1-methyl-3-phenylpentyl)naphthalene. It must, however, always be remembered that the presence of a principal group may override any such considerations.

New class names

Newly invented names fall into two types: those that have been incorporated into large groups such as carbohydrates and steroids whose systematization has been laboriously agreed among the relevant specialists (cf. pp. 118, 119); and those whose systematic equivalents are long or do not clarify the structure at first glance. There is, of course, also the lunatic fringe, which can be exemplified by the name barrelene for (33); this name was given when interaction between the double bonds was postulated and illustrated as in (33); since it has been shown that this effect does not exist, it seems obvious that the not too difficult name bicyclo[2.2.2]octa-2,5,7-triene should be used, but such is the appeal of brevity that the misleading trivial name is still often employed.

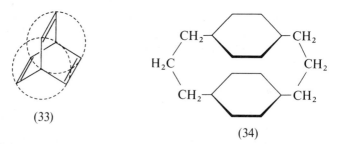

(33)

(34)

An individual name sometimes becomes a class name as analogs are found to form an interesting new set of compounds. Such a case is the cyclophanes, now a generally accepted class name. A typical representative is (34), for which a systematic name would be 1,1′:4,4′-bistrimethylenedibenzene but for which the convenient *p,p′*-[3.3]-

cyclophane was invented. Obvious small changes allow for other linking alkylene groups and for *o*- and *m*-linkage, but differing types of variant can be devised for cases where the benzene rings are replaced by other nuclei, when the alkylene chains contain hetero atoms, where more than two rings are linked in the way shown, or for more than two linking chains. There is already need for systematization of this growing field, but that should be done by the specialists who can assess likely future developments rather than by IUPAC Committees acting alone.

HETEROCYCLIC COMPOUNDS

Of all classes of compound the heterocyclic series presents the greatest variety of structure and thus the most complicated nomenclature. The following is a general outline.

Trivial names

There is a profusion of trivial or semitrivial names, even when the alkaloids are excluded. The IUPAC rules list names of 47 heterocyclic skeletons that may be used in fusion operations (see *Table 5.2*) and of 14 hydrogenated systems that should not be used in fusion operations (see *Table 5.3*). Throughout the heterocyclic series a hetero atom at a ring junction is numbered sequentially as in quinolizine (No. 25 in *Table 5.2*).

Table 5.2 TRIVIAL AND SEMITRIVIAL NAMES OF HETEROCYCLES, IN ORDER OF 'SENIORITY' (see p. 81), WHOSE NAMES ARE APPROVED BY IUPAC FOR USE IN FUSION NOMENCLATURE

1. Thiophene		4. Thianthrene	
2. Benzo[*b*]thiophene (replacing thianaphthene)		5. Furan	
3. Naphtho[2,3-*b*]-thiophene (replacing thiophanthrene)		6. Pyran (2*H*-shown)	

(contd. on p. 82)

Table 5.2 (contd.)

7. Isobenzofuran		17. Pyrimidine	
8. Chromene (2H-shown)		18. Pyridazine	
9. Xanthene*		19. Indolizine	
10. Phenoxathiin		20. Isoindole	
11. 2H-Pyrrole		21. 3H-Indole	
12. Pyrrole		22. Indole	
13. Imidazole		23. 1H-Indazole	
14. Pyrazole		24. Purine*	
15. Pyridine		25. 4H-Quinolizine	
16. Pyrazine		26. Isoquinoline	

* Denotes exceptions to systematic numbering.

Table 5.2 (contd.)

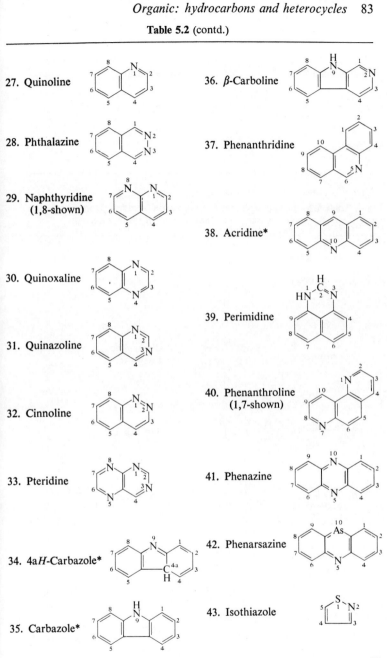

27. Quinoline

28. Phthalazine

29. Naphthyridine (1,8-shown)

30. Quinoxaline

31. Quinazoline

32. Cinnoline

33. Pteridine

34. 4a*H*-Carbazole*

35. Carbazole*

36. β-Carboline

37. Phenanthridine

38. Acridine*

39. Perimidine

40. Phenanthroline (1,7-shown)

41. Phenazine

42. Phenarsazine

43. Isothiazole

* Denotes exceptions to systematic numbering.

(contd. on p. 84)

Table 5.2 (contd.)

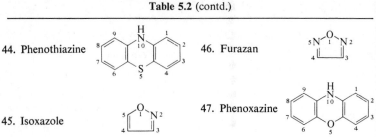

44. Phenothiazine

46. Furazan

45. Isoxazole

47. Phenoxazine

Table 5.3   TRIVIAL AND SEMITRIVIAL NAMES OF REDUCED HETEROCYCLIC SYSTEMS, NOT TO BE USED IN FUSION NOMENCLATURE

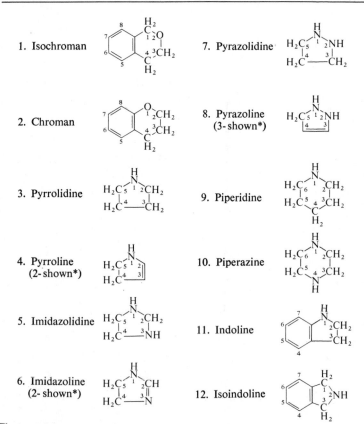

1. Isochroman

2. Chroman

3. Pyrrolidine

4. Pyrroline (2-shown*)

5. Imidazolidine

6. Imidazoline (2-shown*)

7. Pyrazolidine

8. Pyrazoline (3-shown*)

9. Piperidine

10. Piperazine

11. Indoline

12. Isoindoline

* The numeral denotes the position of the double bond.

Table 5.3 (contd.)

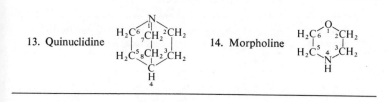

13. Quinuclidine

14. Morpholine

Extended Hantzsch–Widman system

Except for trivial names, this nomenclature has the widest use. Syllables ending in 'a' (elided before a vowel) denote the hetero atoms (see *Table 5.4*) and are followed by other syllables to denote ring size (see *Table 5.5*, p. 86); the latter are in some cases modified to denote the state of reduction of the ring, but in others this always requires hydro prefixes to be used. Most of the suffixes are derived by selection of letters from the appropriate numeral, ir from t*ri* (for three-membered rings), et from t*et*ra, ep from h*ep*ta, oc from *oc*ta, on from n*on*a and ec from d*ec*a; but ole for a five-membered and in for a six-membered ring are from the original Hantzsch–Widman endings[2]. For six- or higher-membered rings a terminal 'e' distinguishes unsaturated nitrogenous rings; -idine is used as ending for fully reduced three- to five-membered rings, and -ane for all fully reduced non-nitrogenous rings; partly reduced four- and five-membered rings

Table 5.4 PREFIXES TO DENOTE HETERO ATOMS IN RINGS

Element	Valence	Prefix	Element	Valence	Prefix
Oxygen	II	oxa	Antimony	III	stiba*
Sulfur	II	thia	Bismuth	III	bismutha
Selenium	II	selena	Silicon	IV	sila
Tellurium	II	tellura	Germanium	IV	germa
Nitrogen	III	aza	Tin	IV	stanna
Phosphorus	III	phospha*	Lead	IV	plumba
Arsenic	III	arsa*	Boron	III	bora
			Mercury	II	mercura

* When immediately followed by '-in' or '-ine', 'phospha-' should be replaced by 'phosphor-', 'arsa-' should be replaced by 'arsen-' and 'stiba-' should be replaced by 'antimon-'. In addition, the saturated six-membered rings corresponding to phosphorin and arsenin are named phosphorinane and arsenane.

Table 5.5 SUFFIXES USED TO DENOTE RING SIZE AND STATE OF REDUCTION OF THE RING IN NOMENCLATURE OF HETEROCYCLES BY THE EXTENDED HANTZSCH–WIDMAN SYSTEM

No. of members in the ring	Rings containing nitrogen Unsaturation*	Saturation	Rings containing no nitrogen Unsaturation*	Saturation
3	-irine	-iridine	-irene	-irane
4	-ete	-etidine	-ete	-etane
5	-ole	-olidine	-ole	-olane
6	-ine†	‡	-in†	-ane§
7	-epine	‡	-epin	-epane
8	-ocine	‡	-ocin	-ocane
9	-onine	‡	-onin	-onane
10	-ecine	‡	-ecin	-ecane

* Corresponding to the maximum number of non-cumulative double bonds, the hetero elements having the normal valences shown in *Table 5.4*.
† For phosphorus, arsenic, antimony, see the special provisions of *Table 5.4*.
‡ Expressed by prefixing 'perhydro' to the name of the corresponding unsaturated compound.
§ Not applicable to silicon, germanium, tin and lead. In this case, 'perhydro-' is prefixed to the name of the corresponding unsaturated compound.

have the following endings (the -oline ending is from Hantzsch–Widman):

No. of members of the partly saturated rings	Rings containing nitrogen	Rings containing no nitrogen
4	-etine	-etine
5	-oline	-olene

Examples:

Δ^3-1,2-Azarsetine*   Silolene

* As exceptions, Greek capital delta (Δ), followed by superscript locant(s), is used to denote a double bond in a compound named according to this system if its name is preceded by locants for hetero atoms.

When more than one hetero atom is present in a ring they are cited in descending Group number of the Periodic Table and in increasing atomic number, e.g. oxathia, thiaza, oxaza (note the elision of 'a' from thiaaza and oxaaza).

Locants precede the names thus produced. Subject to being as low as possible, they must give the lowest permissible number to the hetero atom given first in *Table 5.4* and then to other hetero atoms; they are arranged in the same order as the hetero atoms. As examples, one obtains 1,3-thiazole for (35), 1,2,4-triazine for (36) and 1,2,4-thiazaphosphole for (37) (1,2,4 as a series is lower than the otherwise permissible 1,3,5). (1,3-Thiazole is usually shortened to thiazole.)

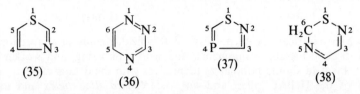

(35)   (36)   (37)   (38)

Indicated hydrogen often requires to be cited as for carbocycles, e.g. 6*H*-1,2,5-thiadiazine (38).

Fusion names for heterocycles

Names of heterocycles of *Table 5.2* or those formed by the Extended Hantzsch–Widman system can be 'fused' with those of hydrocarbons or other heterocyclic systems by the methods outlined above for hydrocarbons. Naturally some special points arise.

The following abbreviations are used in fusion names of heterocycles: furo, imidazo, pyrido, quino, and thieno, but not others.

The hetero atoms in the complete structure receive the lowest numbers permissible when the structure is oriented as for carbocycles; *but* hetero atoms at ring junction (valley) positions are numbered serially without the use of a, b, etc. Formula (39) for pyrano[3,4-*b*]quinolizine illustrates both points.

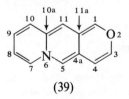

(39)

Square brackets are used as for carbocycles to distinguish locants that refer to the unfused component and not to the final compound. Examples are anthra[2,1-*d*]thiazole (40) and pyrrolo[2,3-*c*]carbazole (41).

Compound (41) serves also to illustrate the loss of two hydrogen atoms (one from the pyrrole and one from the carbazole component) to obtain maximum non-cumulative unsaturation.

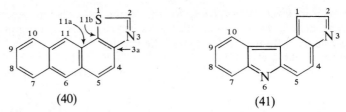

(40)   (41)

Many intricate problems arise in this field, mainly in connection with choosing the components, and with numbering, and workers in this field should familiarize themselves with the details of Section B of the IUPAC rules[1] and with the *Revised Ring Index* and its supplements.

However, there is one welcome simplification, namely when a benzene ring is one fusion component to the Hantzsch–Widman type of name (incidentally a common type of compound): here names such as 1,2-benzoxathiin and 2,1-benzoxathiin are adequate to distinguish (42) from (43), respectively, and there are many similar cases.

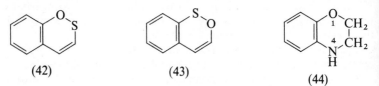

(42)   (43)   (44)

All complete fused structures containing less than the maximum number of double bonds must have 'hydro' prefixes in the name, and these prefixes receive the lowest possible numbers *after* allowance for indicated hydrogen: the name for (34) is 3,4-dihydro-2*H*-1,4-benzoxazine.

Replacement nomenclature

The terms denoting hetero atoms listed in *Table 5.4* for the Extended Hantzsch–Widman names can also be used for general replacement nomenclature: when prefixed to the name of a hydrocarbon they denote replacement of a carbon atom of that compound by the stated hetero atom, and this gives names that often prove most useful, particularly for compounds containing several types of hetero ring members or complex ring systems. An example would be 8*H*-7-thia-1,9-diazaphenanthrene (45). It will be seen that the numbering of the hydrocarbon is retained but, so far as that retention permits, the hetero atoms have lowest numbers. This nomenclature must be applied to the hydrocarbon 'parent', not to a less hetero-substituted system—to avoid plurality of names such as, for instance, 4H-3-thia-7-azaphenanthridine for (45).

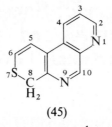

(45)

Bicyclo, tricyclo, etc., names are also a very fruitful field for replacement nomenclature.

Hetero bridges

Hetero bridges, particularly those across aromatic or complex systems, can be usefully named by the 'epi' system. The prefix 'epi' ('ep' before a vowel) is followed by the name of the bridging *radical* (contrast the method for bridging hydrocarbons, p. 78). One thus obtains epoxy for a bridging –O–, epidioxy for –O–O–, epithio for –S–, epimino for –NH–, and so on. Thus compound (46) can be called 9,10-dihydro-9,10-epoxyanthracene and (47) 1,9-dihydro-1,9-epidioxyphenanthrene. Note the dihydro prefixes, as for bridging hydrocarbons, also that epi terms are considered to be part of the main structure and are not alphabeticized among the normal prefixes.

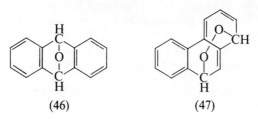

(46)   (47)

REFERENCES

1. *IUPAC Nomenclature of Organic Chemistry, Definitive Rules for: Section A. Hydrocarbons; Section B. Fundamental Heterocyclic Systems; Section C. Characteristic Groups Containing Carbon, Hydrogen, Oxygen, Nitrogen, Halogen, Sulfur, Selenium and/or Tellurium.* 1969. A,B 3rd edn.; C 2nd edn., Butterworths, London (1971). (a) pp. 8–11, 13–15; (b) pp. 22–29; (c) pp. 32–34; (d) pp. 35–37
2. Hantzsch, A. and Weber, J. H., *Ber. deut. Chem. Ges.,* **20**, 3119 (1887); Widman, O., *J. Prakt. Chem.* (2), **38**, 185 (1888)

6 Organic: Some Special Features and Functional Groups

HYDROCARBON AND HETEROCYCLIC RADICALS (SUBSTITUENTS)

Aliphatic

As is well known, univalent radical names derived from alkanes, alkenes or alkynes have the -ane, -ene or -yne ending changed to -yl, -enyl or -ynyl, respectively, e.g. methyl, 2-butenyl. The following specific names may be noted: $CH_2=CH-$ vinyl, $CH_2=CHCH_2-$ allyl, $CH_3CH=CH-$ propenyl.

The change of -ane to -yl for aliphatic radicals applies also to the C_3–C_6 members having an iso-, *sec*- or *tert*- prefix (p. 67) and to neopentane. However, for radicals formed from all other branched hydrocarbons the point of attachment receives number 1, and the most unsaturated, or most substituted, or longest (in that order) chain starting from that atom numbered 1 is regarded as the main (parent) chain; thus for instance, heptane C_7H_{16} gives a radical $CH_3(CH_2)_5CH_2-$ 1-heptyl (or simply heptyl), $CH_3(CH_2)_4CH(CH_3)-$ 1-methylhexyl, and $CH_3(CH_2)_3CH(CH_3)CH_2-$ 2-methylheptyl, etc.; similarly, $CH_3CH_2CH_2CH(C_2H_5)CH_2-$ is 2-ethylpentyl, but $\underset{2}{C}H_3\underset{3}{C}H_2\underset{4}{C}H_2\underset{1}{C}H(CH=CH_2)CH_2-$ is 2-propyl-3-butenyl. Nevertheless, the trivial name isopropenyl for $CH_2=C(CH_3)-$ is a convenient exception.

Names of radicals $CH_3(CH_2)_xCH=$ end in -ylidene (in place of -ane), and those of $CH_3(CH_2)_xC\equiv$ end in ylidyne.

$-CH_2-$ is methylene, and $CH_2=$ has the same name (!); $-CH_2CH_2-$ is ethylene; but names of other saturated $-CH_2(CH_2)_x-$ radicals are built of methylene units, e.g. $-CH_2(CH_2)_3CH_2-$ pentamethylene. Propylene is exceptionally the name for $-CH(CH_3)CH_2-$, leaving trimethylene conveniently for $-CH_2CH_2CH_2-$ (N.B. The hydrocarbon $CH_3CH=CH_2$ is propene.)

Unsaturated aliphatic radicals are named analogously, e.g. $-CH_2CH=CHCH_2-$ 2-butenylene.

There is a relatively new IUPAC rule allowing multivalent aliphatic radicals to be named by *adding* -yl, -ylidene or -ylidyne as appropriate to the name of the hydrocarbon. It gives names such as $\equiv CHCH_2CH_2CH=$ 1-butanyliden-4-ylidyne or $=CHCH=CHCH=$ 2-butenediylidene. Such names are sometimes convenient; thus $-\overset{|}{\underset{|}{C}}-$ is methanetetrayl, so that by the nomenclature for symmetrical compounds (p. 64) $C(C_6H_4COOH)_4$ becomes p, p', p'', p'''-methanetetrayltetrabenzoic acid.

Radicals from cyclic systems

Radicals formed from carbomonocycles are named analogously to those from alkanes, alkenes, etc. For all other radicals formed from cyclic systems there is a general rule that they are named by *adding* -yl, -ylene, -ylidene, -triyl, etc., as appropriate, to the name of the compound (with elision of any terminal 'e'). Examples are indenyl, azulenyl, carbazolyl, pteridinyl and isoxazolyl. There is the following list of exceptions:

Compound	Radical	Compound	Radical
Anthracene	anthryl	Pyridine	pyridyl
Benzene	phenyl	Quinoline	quinolyl
Furan	furyl	Isoquinoline	isoquinolyl
Mesitylene	mesityl	Thiophene	thienyl
Naphthalene	naphthyl	Toluene	tolyl
Phenanthrene	phenanthryl		benzyl
Piperidine	piperidyl	Xylene	xylyl

Additionally, morpholino and piperidino are the radicals formed by removal of H from the N of morpholine and piperidine, respectively, but in analogous cases radical names always end in -yl with a locant corresponding to that of the hetero atom, e.g. 1-indolyl, 1-imidazolyl, 1-pyrrolyl.

Numbering

In all cases, locants for radical endings are as low as permissible by any fixed numbering of the system; they have priority over locants for unsaturation and all substituents. For example, in compound (1) the naphthalene system, being 'senior' to cyclohexene, is named as parent, forcing the latter to be named as substituent, with the numbering shown and giving the name 7-(2-carboxy-6-methyl-5-hexenyl)-2-naphthoic acid.

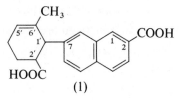

(1)

FREE RADICALS

Free radicals from alkanes or cyclic structures have the same names as for substituent groups, e.g. $CH_3\cdot$ methyl, $C_6H_5\cdot$ phenyl. However, carbene is always used in place of methylene for derivatives of $H_2C{:}$.

The -yl ending is used for free radicals also when the group name may end in 'y', e.g. $CH_3O\cdot$ methoxyl (not methoxy); and amine free radicals also end in -yl, e.g. $CH_3NH\cdot$ methylaminyl, and 1-piperidinyl (not piperidino). Specialists may note $CH_3\dot{N}\cdot$ methylaminylene and its analogs, also hydrazyl for $NH_2NH\cdot$.

IONS

Anions of acids have names formed by replacing -ic acid by -ate, or -ous acid by -ite (see pp. 16, 20).

Anions formed by removing a proton (or protons) from carbon are named by adding -ide (or -diide, etc.) to the name of the parent compound (with elision of 'e' before a vowel). For example, $CH_3CH_2CH_2CH_2^-$ 1-butanide, $C_6H_5^-$ benzenide, and 1,4-dihydro-1,4-naphthalenediide for (2).

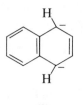

(2)

Cation nomenclature is not simple, except that the ending -ium (with elision of a preceding vowel) is used in all normal cases. We may first note ammonium H_4N^+, sulfonium H_3S^+ and similar hetero cations and their derivatives. Next, for cations derived from compounds by addition of a proton the -ium is added to the systematic or trivial name, with addition of any necessary locants or indicated hydrogen. Typical examples are anilinium $C_6H_5NH_3^+$, 1-methylhydrazinium (3; $N^+ = 1$) and 9aH-quinoxalinium (4). Uronium $OC(NH_2)(NH_3^+) \leftrightarrow HO-C(NH_2)=NH_2^+$ is an exception.

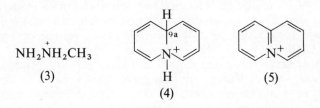

$$NH_2\overset{+}{N}H_2CH_3$$

(3)   (4)   (5)

Cations formally derivable by addition of a proton to an unsaturated compound are named from the latter, e.g. ethenium $CH_3CH_2^+$, benzenium $C_6H_7^+$ and $\overbrace{CH_2=CH=CH_2}^{H^+}$ allenium. Those formally derivable by loss of an electron from a radical are named from the radical, e.g. CH_3CO^+ acetylium, quinazolinylium (5), $C_6H_5^+$ phenylium (but this method affords ethylium for $CH_3CH_2^+$ which duplicates ethenium above).

Radical cations have names ending in -iumyl, e.g. $C_6H_6^{\cdot+}$ benzeniumyl, quinoliniumyl $C_9H_7N^{\cdot+}$, dimethylsulfoniumyl $(CH_3)_2S^{\cdot+}$.

For prefixes, -ium is changed to -ia, which comes in very useful in bicyclo nomenclature, as in (6) 1-ethyl-1-azoniabicyclo[3.3.0]-heptane.

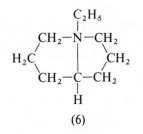

(6)

For further examples, and cases where a single molecule contains both positively and negatively charged centers, the IUPAC rules[1a] should be consulted.

HALO (HALOGENO) COMPOUNDS

Substitutive nomenclature, with fluoro, chloro, bromo and iodo prefixes for halogen substituents, as in 2-chloroquinoline, 2,3-dichloropropionic acid or 1-bromo-6-iodohexane, is widely used and convenient, but for very simple compounds radicofunctional nomenclature, such as methyl iodide CH_3I, benzylidene dichloride (not benzal dichloride) $C_6H_5CHCl_2$ and ethylene dichloride $ClCH_2CH_2Cl$ still has its attractions as emphasizing the main or sole site of reactivity. Also additive nomenclature is sometimes informative when two halogen atoms have been or can be considered to have been added to a double bond, as in styrene dibromide $C_6H_5CHBrCH_2Br$.

When every hydrogen atom of a parent compound is replaced by the *same kind* of halogen atom a prefix 'per-' is used to denote complete replacement; note, however, the difference between perfluoro(decahydro-2-methyl-1-naphthoic acid) $CF_3C_{10}F_{16}COOH$, and hexadecafluorodecahydro-2-methyl-1-naphthoic acid $CH_3C_{10}F_{16}COOH$ as well as decahydro-2-(trifluoromethyl)-1-naphthoic acid $CF_3C_{10}H_{16}COOH$.

Haloform names, e.g. fluoroform, chloroform, for $CH(Hal)_3$, are useful.

$-IO$ is iodosyl, $-IO_2$ iodyl, $-ClO$ chlorosyl, $-ClO_2$ chloryl, $-ClO_3$ perchloryl and $-IX_2$ di-X-iodo.

X_2Hal^+ cations are named as substituted chloronium, bromonium, etc.

$COCl_2$ may be called phosgene, $CSCl_2$ thiophosgene and $C(Hal)_4$ is usually named as a carbon tetrahalide.

ALCOHOLS AND PHENOLS

Alcohols and phenols are named by -ol suffixes if OH is the principal group, otherwise by hydroxy prefixes. Thus names such as 1,8-naphthalenediol and 8-quinolinol are more correct than 1,8-dihydroxynaphthalene and 8-hydroxyquinoline, although both the latter still occur sometimes in the literature.

In aliphatic alcohols the OH groups are cited as -ol when attached to any position in a main chain, e.g. 1-pentanol $CH_3(CH_2)_3CH_2OH$, 1,3-pentanediol $CH_3CH_2CH(OH)CH_2CH_2OH$. Presence of an OH group may determine choice of the main chain (cf. pp. 61, 62).

Radicofunctional nomenclature is still used for simple compounds such as methyl alcohol and benzyl alcohol, notably in the aliphatic and aryl-aliphatic series. For $(CH_3)_3COH$ it is simpler to use *tert*-butyl alcohol than 2-methyl-2-propanol—note that *tert*-butanol is incorrect since there is no *tert*-butane to which a suffix -ol could be added, and it seems a pity that IUPAC did not admit *tert*-butanol to its list of trivial names. The fact that isopropanol is incorrect for $(CH_3)_2CHOH$ since there is no isopropane is less serious because 2-propanol is at least as convenient as isopropyl alcohol.

In such long-known and populated series as alcohols and phenols it is natural that there is a large number of trivial names, far too many to be listed here. A few special points should, however, be noted.

Several alkoxybenzoic acids and alkoxybenzaldehydes have well-known trivial names, e.g. anisic acid $CH_3OC_6H_4COOH$, anisaldehyde $CH_3OC_6H_4CHO$, veratric acid $3,4\text{-}(CH_3O)_2C_6H_3COOH$, vanillin (7) and piperonal (8); to complete these series we require the

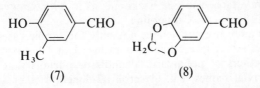

(7)   (8)

alcohols $ArCH_2OH$, where the radical $ArCH_2$ will be named anisyl, veratryl, etc.; this gives, correctly, names anisyl alcohol, etc., to the alcohols; yet when the aryl groups $CH_3OC_6H_4-$, $(CH_3O)_2C_6H_3-$, etc., occur as substituents they also have often been called anisyl, veratryl, etc., which has led to great confusion; it is wiser therefore to name these two types as $ArCH_2-$ alkoxybenzyl and $Ar-$ as alkoxyphenyl.

Trityl is official for the radical $(C_6H_5)_3C-$ as a prefix but not for use in radicofunctional nomenclature, e.g. not trityl alcohol but triphenylmethyl alcohol.

Use of carbinol as a name for alkyl substitution products of CH_3OH, exemplified by dipropylcarbinol for $(C_3H_7)_2CHOH$, is not permitted by the IUPAC rules, but is useful when discussing series of compounds $CHRR'OH$ or $CRR'R''OH$.

Salts of alcohols and phenols should receive the ending -olate, as in sodium methanolate CH_3ONa, sodium phenolate $NaOC_6H_5$ and sodium benzyl alcoholate $C_6H_5CH_2ONa$, but names such as sodium pentyl oxide and particularly the abbreviated methoxide, ethoxide and phenoxide are extremely common; indeed the -anolate ending appears to be to some extent a sop to inorganic nomenclature where it is frequently required for naming coordination complexes.

$(CH_3)_3CONa$ is called sodium *tert*-butyl alcoholate (or oxide) as *tert*-butoxide is incorrect for the same reason as *tert*-butanol; and isopropoxide is correctly 2-propanolate (or 2-propoxide).

ETHERS

For ethers, radicofunctional nomenclature is normal in simple cases, e.g. dimethyl ether, ethyl 1-naphthyl ether, 1-naphthyl propyl ether (note the alphabetical order of the two radicals).

In substitutive nomenclature there is no suffix ending for ether groups, which accordingly must be stated as prefixes.

Prefixes for ether substituents RO– are formed simply by adding 'oxy' to the name of the radical R, e.g. heptyloxy, heptenyloxy; but the yl is elided for the C_1 to C_4 saturated alkyl groups, as in methoxy, and also in the generic name alkoxy.

Alkoxy prefixes are preferred over the ether names when one component is much larger than the other or when several ether groups are attached to one backbone, as in 3β-methoxy-5α-cholestane or 2,3,5-trimethoxyquinoline.

Ether names are convenient not only for simple cases and symmetrical ethers (e.g. dibutyl ether rather than 1-butoxybutane) but also for ethers of polyhydric alcohols or phenols that have well-known trivial names, e.g. glycerol 1,3-dimethyl ether and phloroglucinol trimethyl ether, and in such cases the -ol ending of the alcohol or phenol is best retained.

The prefix for the –O– group is oxy- and for –O–O– is peroxy-.

ROOH are R hydroperoxides and ROOR are R_2 peroxides, where R is the radical name.

ACIDS AND THEIR DERIVATIVES

Acids: general

All organic acids have names ending in -ic acid or -ous acid. There are semitrivial names for many hundreds of acids—saturated, unsaturated, and aromatic, hydroxy and amino carboxylic acids, even for a few sulfur acids; some date back to the 17th century; lists, far from complete, will be found in the IUPAC rules[1b]. Here we shall confine consideration to systematic nomenclature.

A few compounds that have chemically acidic groups other than COOH, SO_3H, etc., have long had acid-type names, and some of these, e.g. picric acid, styphnic acid and ascorbic acid, are too convenient to replace; but others such as cresylic acid (as alternative to cresol) have been eliminated except for crude commercial methylphenols, and the incorrect usage should not be extended. Contrariwise, the amino acids usually have names such as glycine, cysteine, ornithine which emphasize the amino rather than the acid function, though there are also the more chemically acidic glutamic and aspartic acid.

Carboxylic acids

All the usual and some unusual acids occurring in fats or oils, as well as the lower aliphatic acids and some simple aromatic acids, have semitrivial names; from formic to valeric (also isovaleric) acid these are normally used throughout chemistry, but beyond C_5 systematic names are to be preferred for general use; in particular, the names caproic (C_6), caprylic (C_8) and capric (C_{10}) are so similar that they should be avoided.

Systematically, suffixes for COOH are formed in one of two ways: (a) the final 'e' of alkane, alkene or alkyne, etc., is changed to -oic acid, or (b) the COOH group is cited as a suffix -carboxylic acid.

(a) In this nomenclature the C atom of the COOH group is considered as part of the original carbon chain; the ending -oic acid

in fact denotes a grouping $\diagup\!\!\diagdown\!\!\!\!\!\!\!\!$ and not COOH. Thus nonane

$C_8H_{17}CH_3$, for example, leads to nonanoic acid $C_8H_{17}COOH$, and the C of the CH_3 and the COOH receive number 1. For α, ω-dicarboxylic acids the ending -dioic is added to that of the hydrocarbon; nonane gives $HOOC(CH_2)_7COOH$ nonanedioic acid (where no locants are needed).

(b) A suffix -carboxylic acid denotes replacement of H by COOH, i.e. increase in the number of carbon atoms in the molecule. Thus $C_8H_{17}COOH$ might be, but rarely is, called 1-octanecarboxylic acid, and the CH_2 next to the COOH would then be numbered 1.

With these two types of name available, the general rules for principal groups explained on pp. 50, 61 would lead to the name 2-ethylpentanedioic acid (or 2-ethylglutaric acid) for (9); but the carboxylic acid nomenclature for (10), 1,2,4-butanetricarboxylic acid, neatly allows all the COOH groups to be cited in the same way. This is not possible for (11) which has to be named 3-(carboxymethyl)hexanedioic acid.

$$CH_3CH_2CHCH_2CH_2COOH$$
$$|$$
$$COOH$$
$$(9)$$

$$HOOCCH_2CH_2CHCH_2COOH$$
$$|$$
$$COOH$$
$$(10)$$

$$HOOCCH_2CH_2CHCH_2COOH$$
$$|$$
$$CH_2COOH$$
$$(11)$$

For COOH groups attached to cyclic structures the carboxylic acid nomenclature (or a semitrivial name) is obligatory, e.g. 1-pyrrolecarboxylic acid, 2,2′-biphenyldicarboxylic acid, as there is no CH_3 for conversion into COOH.

-oic/-ic

The reader will be familiar with many acid names ending simply in -ic acid, not -oic or -carboxylic. These include all the trivial (semitrivial) names of aliphatic acids but not their systematic equivalents. Among cyclic acids there is no regularity, e.g. benzoic, toluic, naphthoic, phthalic, cinnamic, furoic, nicotinic.

Other acids

The sulfur acids, RSO_3H (sulfonic), RSO_2H (sulfinic) and RSOH (sulfenic) are important compounds. Their terminations are treated in the same ways as carboxylic for formation of derivatives. The R groups are cited by the name of the parent compound RH, giving RSO_3H benzenesulfonic acid, etc. (not the radical form phenylsulfonic). Other sulfonic acids, as well as the alternative sulfane names for sulfenic acids, are discussed in a separate section (pp. 109–112).

For phosphorus acids see p. 118.

Prefixes

Prefixes for the acid groups are carboxy HOOC–, sulfo HO_3S–, sulfino HO_2S– and sulfeno HOS–.

Acyl radicals RC(O)– can be named from the acid by changing -oic acid or -ic acid to -oyl or by changing -carboxylic acid to -carbonyl; but the 'o' of oyl is omitted for aliphatic acid radicals containing less than six carbon atoms. Examples are hexanoyl-, benzoyl-, phthaloyl- [$C_6H_4(CO-)_2$], cyclohexanecarbonyl- and acetyl-.

One meets here a difficulty. The cyclohexanecarbonyl type of name for acyl radicals is used only in radicofunctional nomenclature, e.g. for cyclohexanecarbonyl chloride. If the acyl group is to be named as a substituent that type of name is replaced by a compound prefix for R–CO, where R is the radical name and CO carbonyl (as usual); thus one arrives at, for example, 1-(cyclohexylcarbonyl)-piperidine.

The dual type of nomenclature also applies to the sulfur acids; for the compound prefixes for $X-SO_2-$, $X-SO-$ and $X-S-$, sulfonyl is used for $>SO_2$, sulfinyl for $>SO$ and thio for $>S$; thus $C_6H_5SO_2-$, for example, is phenylsulfonyl when operating as a substituent.

Similar compound prefixes are used in many other cases. Very important is the ester prefix $ROOC-$, clearly named as alkoxycarbonyl-, e.g. methoxycarbonyl-, benzyloxycarbonyl- but $ClOC-$ is chloroformyl and $NH_2C(O)-$ has a special name carbamoyl, since NH_2COOH is carbamic acid.

Ions

Anions $RCOO^-$ and the analogous sulfur anions are named by changing -ic acid to -ate, giving an ending -ate or -oate according to whether the acid name ends in -ic or -oic. Examples are sodium acetate CH_3COONa, potassium benzenesulfonate $C_6H_5SO_3K$, potassium decanoate and triammonium 1,2,4-benzenetricarboxylate.

Salts of amino acids having trivial names are best named by periphrasis such as glycine sodium salt, although names of the type sodium glycinate are common and not so obviously wrong in the light of modern inorganic nomenclature.

The prefix for $^-OOC-$ is carboxylato.

Esters

By an analogy with salts that has been known to be wrong for getting on for 100 years, esters are named analogously to salts, the radical name for the R of $RO-C(O)-$ replacing the name of the cation of a salt, as in methyl acetate.

Prefixes for ester groups were mentioned above.

Amides

For compounds $RCONH_2$, RSO_2NH_2, etc., the ending -ic, -oic or -ylic acid is changed to -amide, giving acetamide, hexanamide, 1-pyrrolecarboxamide, benzenesulfonamide, and so on. When the N is substituted, giving $RCONHR'$, $RSO_2NR'_2$, etc., the compound can be named as an *N*-substituted amide, e.g. *N*-methylacetamide, *N,N*-diethylbenzamide. There is an alternative of citing the RCO group as an acyl radical, and this is done when the R′ group on the nitrogen is more complex than the R group of the acyl; a simple case is 1-acetylpiperidine. In many cases choice between the two methods is optional.

With a few simple R′ groups the process is simplified; if R′ is a simple aryl group, names such as acetanilide, sulfanilanilide and formo-*p*-toluidide may be used.

The prefix for NH_2CO- has been mentioned on p. 99.

Monoamides of common dicarboxylic acids may be given names ending in -amic acid or -anilic acid, e.g. phthalamic acid $NH_2COC_6H_4COOH$, succinanilic acid $C_6H_5NHCOCH_2CH_2COOH$.

Various situations of further complexity arise or can be thought up for acyl derivatives of amines, and some of these are discussed in the IUPAC rules[1c].

Derivatives of dibasic acids

In names of derivatives of dibasic acids it is customarily assumed (probably unwisely) that both acid groups are involved. Thus sodium succinate denotes the disodium salt, malononitrile is $CH_2(CN)_2$ and phthalamide is the diamide. Monometal salts are treated as acid salts, e.g. sodium hydrogen succinate, and esters similarly; but for many types of derivative one group must be cited as prefix, as in *p*-(chloroformyl)benzoic acid $ClCOC_6H_4COOH$.

LACTONES, LACTIMS AND THEIR SULFUR ANALOGS

There are too many ways of naming these compounds that have caused much trouble, so here we shall outline only the methods that have wide use and IUPAC approval.

Many lactones have trivial names, notably coumarin (12), isocoumarin (13) and phthalide (14).

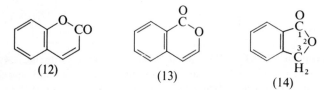

Lactones derivable from monohydroxy acids may be named as lactones or by an -olide ending; compound (15) can be called δ-valerolactone or 5-pentanolide.

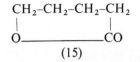

$$CH_2-CH_2-CH_2-CH_2$$
$$O\underline{\hspace{3cm}}CO$$

(15)

More complex, particularly polycyclic, compounds may be named as carbolactones, this suffix denoting addition of a –CO–O– group with the CO taking the lower locant; for example, compound (16) would be 4,3-pyrenecarbolactone and (15) could be 1,4-butanolactone.

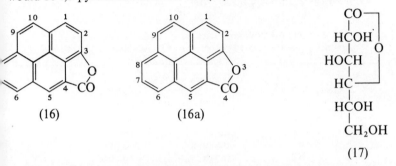

(16)   (16a)   (17)

When a polyhydroxy acid has a trivial name, a derived lactone name may be patterned on D-glucono-1,4-lactone for (17).

Compounds that have a group –CO–NH– or –C(OH)=N– as part of a ring are named on the olide principle but with lactam or lactim, respectively, as suffix.

Sultones are internal anhydrides in which SO_2 replaces the CO of a lactone and are named on the carbolactone principle but with sultone replacing carbolactone as suffix to the parent name. Sultams are the SO_2 analogs of lactams and have a sultam suffix.

All these compounds may, of course, be also named as heterocycles by the ordinary methods, e.g. (15) as tetrahydro-2-pyrone and (16; renumbered as 16a) as pyreno[3,4-*bc*]furan-4-one.

NITRILES AND DERIVED GROUPS

By one method, confined to the aliphatic series, nitriles can be named by adding nitrile, to signify ≡N, to the name of the hydrocarbon with the same number of carbon atoms; this is analogous to the -oic acid nomenclature and gives, for example, nonanenitrile $C_8H_{17}CN$ and nonanedinitrile $NC(CH_2)_7CN$.

The carboxylic acid nomenclature is mirrored in the ending carbonitrile for the –CN group, as in 1-pyrrolecarbonitrile C_4H_4N–CN and 1,2,4-butanetricarbonitrile [analogous to the acid (10)].

When the corresponding acid has a trivial name, the ending -ic acid or -oic acid is changed to -onitrile, as in acetonitrile and benzonitrile.

The prefix for a NC– group is cyano-.

Radicofunctional names, with cyanide to denote the –CN function, are normally used only in special and simple cases such as benzoyl cyanide C_6H_5COCN. However, it is universal with the related groups shown in *Table 6.1*.

For certain dinitriles see p. 101.

Table 6.1   CYANIDE AND RELATED GROUPS IN ORDER OF DECREASING PRIORITY FOR CITATION AS FUNCTIONAL CLASS NAME

Group X in RX	Functional class ending and generic name of class	Prefix
–CN	cyanide	cyano-
–NC	isocyanide*	isocyano-
–OCN	cyanate	cyanato-
–NCO	isocyanate	isocyanato-
–ONC	fulminate	—
–SCN	thiocyanate	thiocyanato-
–NCS	isothiocyanate	isothiocyanato-
–SeCN	selenocyanate	selenocyanato-
–NCSe	isoselenocyanate	isoselenocyanato-

* Not isonitrile or carbylamine.

ALDEHYDES

Aldehydes are named (i) by a suffix -al to the name of an aliphatic hydrocarbon (with the usual elision of 'e') to denote a grouping $\diagdown\!\!\!\!\diagup\!\!\!\!{}^{H}_{O}$, (ii) by adding -carbaldehyde to the name of a hydrocarbon (aliphatic or cyclic) or heterocycle to denote the group –CHO, or (iii) by changing the ending -oic acid or -ic acid of a semitrivial acid name to aldehyde. The procedures (i) and (ii) are as for acids and need no further explanation; procedure (iii) gives names such as acetaldehyde and benzaldehyde.

A few aldehydes have specific trivial (e.g. vanillin) or semitrivial (e.g. piperonal) names.

Oximes, semicarbazones, hydrazones and similar derivatives of aldehydes are named by citing the name of the derivative as a separate word, as in nonanal semicarbazone and benzaldehyde 2,4-dinitro-phenylhydrazone. Occasional abbreviations such as acetaldoxime and benzaldoxime are still met but seem unnecessary, and analogs should not be proliferated.

Some compounds containing both aldehyde and carboxyl groups have names of the type illustrated by succinaldehydic acid (cf. -amic acids, p. 100).

Formyl- is the name for the OHC– prefix, this being also the acyl radical of formic acid.

KETONES

In substitutive names of ketones the =O group is cited as suffix -one or prefix oxo- according to whether it does or does not signify the principal group; examples are 2-butanone $CH_3CH_2COCH_3$ or 4-oxocyclohexanecarboxylic acid.

In radicofunctional nomenclature the $\diagdown C{=}O$ group is the function, and the class name is ketone, which is preceded by separate words specifying the radicals R and R′ of RR′CO; thus $CH_3CH_2COCH_3$ becomes ethyl methyl ketone (note the alphabetical order, ethyl methyl); the prefix is oxo-, as in substitutive nomenclature.

Trivial and semitrivial names are not quite so common as in some other series, but several are important; acetone, CH_3COCH_3, benzil $C_6H_5COCOC_6H_5$, deoxybenzoin (18), chalcone (19), and ketene $CH_2{=}CO$ spring to mind [note the locants in (18) and (19)]; benzil has been made the pattern also for others, such as furil $[(2\text{-}C_4H_4O)CO]_2$ where C_4H_4O is furyl, and there is benzoin $C_6H_5CH(OH)COC_6H_5$ and its analogs.

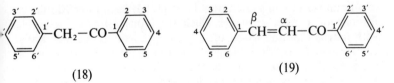

$$\qquad\qquad (18) \qquad\qquad\qquad\qquad (19)$$

So far so simple, but there are complications to follow.

First, it will be obvious that RCO–, being an acyl group, may be named as such when substituted into the other group R′, and this is customary when R′ is large and the acyl group relatively small; indeed, as an example, 2-acetylpyrene seems the best name for the ketone (20). There is a long-standing variant of this whereby for benzene and naphthalene derivatives the -yl of a small acyl group is changed to 'o' and phenone and naphthone are used for the ring system into which the acyl group is substituted; as familiar examples we may cite acetophenone $C_6H_5COCH_3$, benzophenone $C_6H_5COC_6H_5$ and 2'-propionaphthone for (21) [note the priming (dashes) of the ring locants].

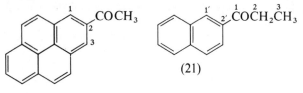

(20)   (21)

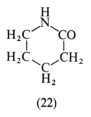

(22)

It is, however, cyclic ketones that require the most care. To start with, a CO group next to O or N characterizes, respectively, a lactone or a lactam and such compounds can be named as belonging to those classes; in fact, many lactones and most lactams are more usually named as ketone derivatives of a non-ketonic parent; the lactam (22) is normally called 2-piperidone [contrast the lactone (15)].

It was explained in Chapter 5 that xH (where x is locant and H is termed indicated hydrogen) is placed before the name of a cyclic compound when an extra hydrogen atom is required after the maximum number of conjugated double bonds has been inserted into the formula of a cyclic structure. Such situations arise, for example, with indene (23) and carbazole (24); both these compounds are so common that the $1H$ of indene (23) and the $9H$ of carbazole (24) are normally omitted from the names; however, the isomers (23a) and (24a) would be named $4H$-indene and $1H$-carbazole, respectively, and would cease to be hypothetical if the H_2 were replaced by, say, two methyl groups.

Now since a suffix -one denotes replacement of 2H by =O, the ketone (25) becomes 1-indenone, and (26) becomes 4-indenone. Here the indicated hydrogen prefix may be regarded as unnecessary in view of the locant of the CO group; but *Chemical Abstracts* inserts indicated hydrogen into the name of the ketone when it is used in

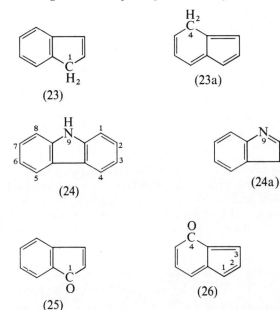

(23)   (23a)

(24)   (24a)

(25)   (26)

the parent hydrocarbon or heterocycle, but not otherwise (this practice makes their indexing more logical); thus since indene (without 1*H*) suffices, so does indenone, but since (23a) is written 4*H*-indene, therefore (26) is named 4*H*-inden-4-one.

A different trouble meets us if we introduce a keto group into a parent ring system that does *not* require extra hydrogen, say pyrene; in such cases extra hydrogen must be added to the *ketone* skeleton, giving, say, (27) or (28), and the position of this hydrogen must be indicated in the name, by means of *H*. But, as this hydrogen arises owing to introduction of the keto group and is not due to the pyrene skeleton itself, the *H* in the name should be associated with the -one suffix and not with the name pyrene. This is done as in 2(1*H*)-pyrenone (27) and 2(7*H*)-pyrenone (28).

Similar situations arise, of course, with heterocyclic ketones. The

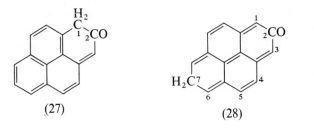

(27)   (28)

ketone (29) requires the name 2*H*-carbazol-2-one, but the ketone (30) must be named 3(2*H*)-phenanthridinone.

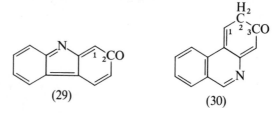

(29)   (30)

A small digression must be made before proceeding further. Some ketone names are abbreviated; common cases are anthrone (for 9-anthracenone) and omission of 'in' in acridone, pyridone, piperidone, quinolone and isoquinolone; in all these cases the indicated hydrogen is also omitted. The long established 4- (31) and 5-pyrazolone (32) do not conform to the rules since two H can be removed from each ring, but in spite of this these names are widely used; the simplified names 4-oxazolone, 4-isoxazolone and 4-thiazolone signify compounds with CO–CH$_2$–O or CO–CH$_2$–S sequences in the ring.

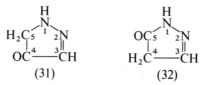

(31)   (32)

Reverting to the general procedure we must note the nomenclature of certain cyclic di- or poly-ketones that can cause trouble to authors. Pyridine derivatives can be used as examples. Compound (33) is 2(1*H*)-pyridone, though the 1*H* is usually omitted; no extra hydrogen is needed in (34) which is simply 2,3-pyridinedione and (35) is similarly 2,5-pyridinedione; (36) (also a diketone) is 2(1*H*),4(3*H*)-pyridinedione and (37) could be 2(1*H*), 4(3*H*), 6(5*H*)-pyridinetrione, but since the saturated heterocycle has the name piperidine it is simpler to name ketone (37) as 2,4,6-piperidinetrione. Note also that (38) becomes 3,4-dihydro-2(1*H*)-pyridone.

The anomalies are all too obvious: the series (33) to (36) are all named as pyridine derivatives, although (33), (34) and (35) are derivable from dihydropyridines and (36) is derivable from a tetrahydropyridine; (38) receives a dihydro prefix although it is derivable from the same tetrahydropyridine. The state of hydrogenation, from a chemical viewpoint, is not considered in this nomenclature—it is solely a matter of whether extra hydrogen is needed after first the

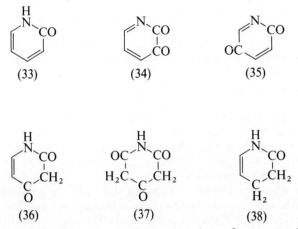

(33)   (34)   (35)

(36)   (37)   (38)

keto group and then the maximum number of non-cumulative double bonds have been introduced into the ring.

Now, in the past, nomenclature has been used which does take regard to the chemical state of hydrogenation of the ring, namely, that for instance (33) would be named 1,2-dihydro-2-oxopyridine. But this hydro-oxo system has a fundamental failing, namely that it does not provide a suffix or functional class name for the ketone group that is the principal group in these cyclic ketones. The hydro-oxo names thus have no place in modern nomenclature, and in spite of the apparent chemical anomalies the system outlined above, with *H* where necessary, should be used today.

Derivatives

Oximes, hydroazones, etc., are named as for aldehydes (see p. 102).

AMINES

Amines have names ending in -amine or, for compounds containing nitrogen as component of a cyclic skeleton, one of the syllables listed for them in *Table 5.5* (p. 86). The ending amine may be used (a) as a suffix in substitutive nomenclature, as in ethanamine $CH_3CH_2NH_2$ or (b) by the older method of attaching it after the name of the radicals replacing the H of NH_3, as in ethylamine, ethylmethylamine, etc.

Method (a) is particularly used for aliphatic polyamines such as 1,3,5-pentanetriamine (simpler than 1,3,5-pentanetriyltriamine) and when amino groups are attached directly to a polycarbocyclic or heterocyclic system, as in 1-pyrenamine or 2-pyridinamine.

Method (b) is widely used for simple compounds, particularly in the aliphatic series and in benzylamine and naphthylamine ($C_6H_5NH_2$ has, of course, the semitrivial name aniline). This usage is in fact a left-over from Hoffmann's contribution to the theory of types.

The prefix for H_2N- in nomenclature is amino-, but the free radical $H_2N\cdot$ is aminyl.

There are many semitrivial names, notably for simple benzene derivatives (toluidine, xylidine, etc., as well as aniline), also for alkaloids and amino acids.

For secondary and tertiary amines with different groups attached to the same N atom the largest group is usually considered to provide a parent amine, and the other groups are then named as substituents, preferably preceded by a locant $N-$ or $N,N-$ which, however, is omitted in simple cases; names such as N,N-diethylbenzylamine, dimethyldodecylamine, methylaniline and N^2-ethyl-2-pyridinamine are common. This procedure also applies to heterocyclic compounds containing an NH group, the numerical locant then replacing N, as in 1-phenylimidazole.

Amine salts

Amine salts, being derivatives of quadricovalent nitrogen, should systematically be named as -onium or -inium salts, for example $[NH_2(C_2H_5)_2]^+Cl^-$ diethylammonium chloride, $[C_6H_5NH_3]^+Cl^-$ anilinium chloride, $[C_5H_6N]^+[C_6H_2O_7N_3]^-$ pyridinium picrate. Occasional use is, however, still made of names such as aniline hydrochloride, pyridine picrate (note that 'hydro' is included only with the halide names: aniline hydrochloride, hydrobromide or hydriodide, but aniline picrate or nitrate; this old, irrational practice is probably derived from the names for the acids). Such 'irregular' names for amine salts are, however, necessary for unsymmetrical bases when the point of attachment of the proton cannot be defined, as in 1,6-naphthalenediamine monohydrochloride (if the point of attachment were known, the specific name would be, e.g. 6-amino-1-naphthylammonium chloride).

Care must be taken with amine salts of dibasic acids; for example, $[C_6H_5NH_3]^+[HSO_4]^-$ is anilinium hydrogen sulfate and $2[C_6H_5NH_3]^+[SO_4]^{2-}$ is dianilinium sulfate, whereas aniline sulfate or anilinium sulfate might refer to either salt.

Quaternary salts have systematic names such as triethylmethylammonium chloride, N,N,N-trimethylanilinium bromide and

1-methylquinolinium picrate. Quaternary salts are, however, often handled by an older method, as metho-salts, e.g. triethylamine methochloride, quinoline methopicrate: this older method should not normally be used, though it is useful for compounds of uncertain composition (cf. hydrochloride, etc.). Salts derived by addition of $(CH_3)_2SO_4$ are named after the pattern of 1-methylpyridinium methyl sulphate or 1,6-naphthalenediamine monomethosulphate,

the trivial name methosulphate denoting $[{>}NCH_3]^+[CH_3SO_4]^-$.

The prefix azonia for the N^{IV} cation and its use have been mentioned on p. 93.

The IUPAC rules should be consulted for a very detailed account of the nomenclature of the various types of compound containing more than one nitrogen atom[1d].

SULFUR COMPOUNDS

Sulfonic, sulfinic and sulfenic acids have been mentioned on p. 98, but there are many further aspects of organic sulfur compounds for which IUPAC nomenclature is complex, often little understood and generally unsatisfactory. Consider the following bare facts.

The syllables 'thio', to denote sulfur, appear in thiol, thione, thionium, thionio, thionia and as thio itself; and thia- as already mentioned, denotes a sulfur atom in a ring on the Hantzsch–Widman system (p. 85) or replacement of carbon by sulfur in replacement nomenclature (p. 88).

Thiol, as a suffix, denotes –SH present as principal group in substitutive nomenclature (*Table 4.3*), as in methanethiol CH_3SH; but mercapto is used for HS– as prefix.

Thione can be used to denote =S doubly bonded to one atom when it is the principal group, just as -one denotes carbonyl =O, e.g. 2-butanethione for $CH_3CH_2C(=S)CH_3$. Nevertheless, the generic name for such compounds is thioketones and the prefix for this =S is thioxo!

Thionium denotes a cation $[S^{III}]^+$, as in trimethylsulfonium chloride $[(CH_3)_3S]^+Cl^-$. Thionio is used when R_2S^+- is named as a prefix. Thionia- is the prefix for an S^{III} cation in replacement nomenclature (p. 93).

Thio- (unmodified) requires more consideration. It is used in two distinct ways: to denote –S– as a linear component and to replace oxygen. (It is the latter that leads to thioketone and thioxo-.)

Linking the divalent thio –S– to one organic group gives the radical RS–, which is named R-thio-, e.g. methylthio- for CH_3S–, in the same way as the divalent radical $\diagdown CO$ carbonyl leads to methoxy-carbonyl for CH_3OOC–. Radical names of methylthio- type are placed as prefix when this group is attached to a large group, as in 1-(methylthio)pyrene. There is naturally a radicofunctional alternative to this, namely the use of sulfide as the functional class name (cf. inorganic chemistry which led to this older nomenclature), preceded by the radicals of RR′S as separate words, as in ethyl methyl sulfide $CH_3CH_2SCH_3$ or dimethyl sulfide $(CH_3)_2S$; such names are still used for simple compounds.

Salts of thiols are thiolates, e.g. CH_3CH_2SNa sodium ethane-thiolate or, by radicofunctional nomenclature, sodium ethyl sulfide.

It will then be apparent that, if AlkS– is alkylthio-, then an HS– group could have been called hydrothio-; but this was rejected in favor of mercapto-; yet mercapto- was derived from mercaptan for RSH, an old name that was itself rejected in favor of thiol!

A second type of use of thio for –S– is in naming symmetrical compounds containing a functional group that is preferred to –S–; thus $S(CH_2CH_2COOH)_2$ is simply named as 3,3′-thiodipropionic acid.

More diverse are the applications of thio to denote replacement of oxygen, in effect for naming sulfur analogs of oxygen compounds. Thio indicates replacement of oxygen singly bonded to two groups or atoms in names such as thiophenol C_6H_5SH and thiocresol $CH_3C_6H_4SH$; also in heterocyclic compounds such as 4H-thiopyran (see below), thiochroman, thiochromone and the like.

Confusion can thus arise when more than one oxygen atom is present in a molecule; we shall discuss thiocarboxylic acids below, but two special cases should first be noted. Compound (39) is 4H-pyran, so with thio replacing –O– the compound (40) becomes

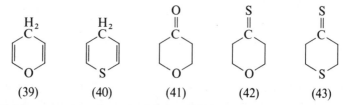

(39)   (40)   (41)   (42)   (43)

4H-thiopyran; but since the ketone (41) is 4-pyrone, the analog (42) becomes 4-thiopyrone, and then how is one to name the thioketone (43)? Since thiopyran and pyrone are well established it seems better to abandon abbreviation and use thiopyran-4-thione for (43). It should be noted that a different solution, namely thiapyran for (40) had been used earlier, but this was rejected by IUPAC as opening the door too wide to replacement of other hetero atoms by sulfur, which might lead to $4H$-thiapyridine as an alternative for thiopyran (40).

The second special case is thioglycolic acid which is often used for $HSCH_2COOH$ as the sulfur analog of glycolic acid $HOCH_2COOH$; the difficulty is that thioglycolic acid is also the name of a glycolic acid in which an oxygen atom of the COOH group has been replaced by sulfur. Here the solution is to use the equally correct mercaptoacetic acid for $HSCH_2COOH$.

There are analogs for both these special cases.

Thio and dithio are also widely used in names of thio acids, both inorganic (see p. 16) and organic, such as thioacetic acid $CH_3C(OS)H$ and dithiobenzoic acid $C_6H_5CS_2H$ to denote replacement of one or both of the oxygen atoms of the COOH group. The dithio name is unambiguous; the monothio name does not specify whether the $=O$ is converted into $=S$ or the OH into SH, and precisely this indecision is intended because it is normally not known whether the H is attached to O or S in the thio acid (though presumably more to S than to O). If this distinction is known, the species are to be called, for example, thioacetic S-acid for $CH_3C(=O)SH$ and thioacetic O-acid for $CH_3C(=S)OH$. The distinction is, however, usually known for esters, and here it is easily effected as S-methyl thioacetate $CH_3C(=O)SCH_3$ and O-methyl thioacetate $CH_3C(=S)OCH_3$. Previous distinction by means of the endings thiolic, thionic and thionothiolic has been discarded by IUPAC. Trithio may be needed for, e.g. phosphoric acids and carbonic acid, but solutions similar to the above are then also available.

The situation can be equally complicated for thio sulfur acids; ethanedithiosulfonic acid is $C_2H_5S(S_2O)H$, for example, where attachment of the H is not specified, and we have names such as benzenetrithiosulfonic acid $C_6H_5S(S_3)H$ and benzenedithiosulfinic acid $C_6H_5S(S_2)H$. Finally on this subject we have carbothioic and carbodithioic acid for the sulfur analogs of acids named as carboxylic acids.

The names R-sulfonyl and R-sulfinyl for the radicals RSO_2- and $RSO-$ (R is the radical name), respectively, have been mentioned on p. 99. These should be contrasted with sulfone as the radicofunctional class name for an $\diagup\!\!\!\!\diagdown SO_2$ group, as in $CH_3CH_2S(O_2)CH_3$ ethyl methyl sulfone, and sulfoxide correspondingly for an $\diagup\!\!\!\!\diagdown SO$ group.

As prefix names for RS_2-, RS_3-, etc., there are the names R-dithio-, R-trithio-, etc. (R is the radical name); disulfide, trisulfide, etc., then replace sulfide in radicofunctional names such as ethyl methyl disulfide $CH_3CH_2S_2CH_3$. Recently a new nomenclature has been developed for compounds known or believed to have chains of sulfur atoms. HSH is termed sulfane, HSSH is disulfane, HSSSH is trisulfane, and so on. Groups or atoms replacing the hydrogen are cited as prefixes. Thus C_6H_5SSH is phenyldisulfane and CH_3CH_2-SCH_3 can be called ethylmethylsulfane; a prefix is used even if the compound is acidic, as CH_3SSSOH which is called 1-hydroxy-3-methyltrisulfane; the salt CH_3SSSNa is sodium methyltrisulfanide and a salt $CH_3SSSONa$ is sodium 3-methyl-1-oxidotrisulfane (oxido is O^-).

INDICATED HYDROGEN—GENERALIZATION OF THE CONCEPT

Indicated hydrogen, symbolized by *H* preceded by a locant, has been discussed in the pages above in two connections. One (p. 73) was in names such as 1*H*-fluorene or 4*H*-pyran, where after introduction of as many non-cumulative double bonds as possible into a ring system, there remains 'extra' hydrogen to be introduced and there is more than one possible position for that hydrogen, so that its actual position must be 'indicated' by a locant. The other occasion (p. 104) was with ketones; here one met ketones formally derived by replacement of CH_2 by CO in a ring system already carrying an indicated hydrogen, so that the keto group is named merely by a suffix -one, as in 1*H*-fluorenone; but cases were also found where indicated hydrogen was required after, but not before, the keto group was placed on the ring skeleton; the latter case is exemplified by naphthalene (44) and 1(2*H*)-naphthalenone (45) where the symbolism 1(2*H*) shows that the 2*H*-hydrogen is present because of the 1-keto group.

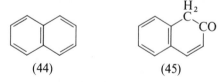

(44)   (45)

Though simple cases such as those discussed above and previously are the most often encountered, complications and extension of these procedures are not infrequent.

As complications of the ring structure alone, we can note the spiran

(46) which is named spiro[2*H*-indene-2,1'-cyclopentane] (so as to pin down the isomeric indene skeleton); also the radical (47) which is 2(1*H*)-naphthylidene, the symbolism here being 2(1*H*) to show that the need for the extra 1*H*-hydrogen arises from introduction of the -ylidene radical.

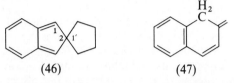

(46)   (47)

Compound (48) is analogous to the simple ketones and is named as a derivative of 2*H*-carbazole—simply 2*H*-carbazole-2,2-dicarboxylic acid. The general case corresponding to the ketone (45) arises when *any* principal group is introduced where there was no hydrogen in the parent compound, or only insufficient hydrogen. For example, (49) is 2,2(1*H*)-naphthalenedicarboxylic acid, the isoquinoline derivative (50) is 2(1*H*)-isoquinolinecarboxylic acid and (51) is 8a(1*H*)-isoquinolinecarbonitrile.

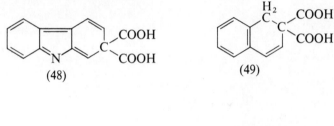

(48)   (49)

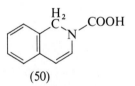

(50)   (51)

Perhaps the most perplexing feature is that this 2(1*H*) type of procedure is applied only to principal groups, i.e. when the introduced group is named as a suffix. A compound (52) is called 1,2-dihydro-2-methylisoquinoline, in contrast to (50). The difference is due to the concept of 'detachable' groups: prefixes (not suffixes) denoting substituents and hydro are detachable; others (cyclo, benzo, oxa, iso, indicated hydrogen and a great many others) are non-detachable. Non-detachable prefixes, like suffixes, must remain attached as closely as possible to the parent name, even in indexes;

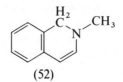

(52)

detachable prefixes can be rearranged among themselves in alphabetical order (see p. 55) and can be placed after the parent name in indexes. Thus (51) must be indexed as 8a(1*H*)-Quinolinecarbonitrile, but (52) can be indexed as Quinoline, 1,2-dihydro-2-methyl-. These indexing procedures have as their very desirable object to keep both entries beginning with Quinoline; it may appear here that indexing is the tail wagging the nomenclature dog, but in fact no nomenclature is good that does not lend itself readily to sound indexing.

Finally we may note that indicated hydrogen has the highest priority for lowest locant (cf. p. 56), higher even than the position of

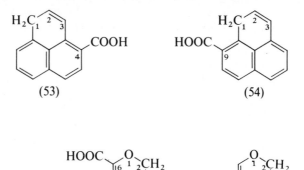

(53)   (54)

(55)

(56)

a radical valence or of a principal group. Compound (53) is 1*H*-phenalene-4-carboxylic acid and its isomer (54) is 1*H*-phenalene-9-carboxylic acid; compound (55) is 2*H*-pyran-6-carboxylic acid and its isomer (56) is 2*H*-pyran-4-carboxylic acid; i.e. the pairs of isomers are treated as different acids derived from the same (non-acidic) parent, not as different constitutional isomers of the same acid. The reason is that for (53/54) there can be only two parent phenalenes—1*H*- and 2*H*-phenalene—so that it would seem wrong to call (54) 3*H*-phenalene-4-carboxylic acid; and analogously for (55/56).

There are other problems for indicated hydrogen, but perhaps enough has been said to show the depth of these waters.

STEREOISOMERISM

Definition of stereochemical terms is more a matter of chemistry than of nomenclature and will not be attempted here. However, inorganic and organic chemists equally with biochemists require to describe in detail the steric nature of their compounds and reactions, and this generality requires a nomenclature agreed overall; that, however, is prevented from completion by the very rapid progress being made in both theoretical and experimental aspects and by the different approaches of the various investigators. It may thus be useful to review here very briefly the present situation and some of the key matters involved.

Many of the older terms remain in use; a few new ones have been generally adopted and others are likely to be so adopted soon. The modern textbooks and reviews go their own way less in stereo-chemical terminology than in general nomenclature, and it is hoped that recent Tentative Rules for 'Fundamental Stereo-chemistry' published by IUPAC[2] will prove valuable on the less recondite matters.

The term 'structure' is allotted very wide application—to any aspect of the organization of matter. 'Constitution' is the term to be used to denote the nature and sequence of bonding of atoms in a compound. Thus 'constitutional isomerism' is correct, and 'structural isomerism' is incorrect in designation of stereochemistry.

Stereoisomers are isomers differing only in the arrangement of their atoms in space. Those that are interconvertible by rotation of atoms or groups around a single bond (or, in view of some workers, also around multiple bonds) are termed conformational isomers, and those that are conceived as requiring bond-breaking for intercon-version are termed configurational; this loose wording is here intentional, for it has proved impossible so far, in spite of many attempts, to find a generally acceptable method of drawing a sharp boundary between these two, generally distinguishable, concepts; the main features of the problem are discussed in Appendix 1 to the IUPAC Tentative Rules[2].

The word 'chirality' has come into general use, meaning 'handed-ness', the property of non-superposability on the mirror image, i.e. the criterion for the existence of enantiomers. The adjective 'chiral' is applied to optically active molecules and is also used in 'chiral center' as an abbreviation for 'center of chirality' (similarly chiral axis and chiral plane).

For organic compounds in general, the sequence-rule symbols *R*, *S*, *r*, *s*, *M* and *P* are now widely used; simple aspects of their assignment are described in most recent textbooks, but the original paper[3] should be consulted for full detail; there are two useful summaries[4] for the general worker.

Prefixes *cis-* and *trans-*, to describe isomerism around double bonds or, more generally for four coplanar ligands, are still used but they are being superseded by *Z-* (German, *zusammen* = together = *cis*) and *E-* (German, entgegen = opposite to = *trans*)[5]; being based on the sequence rule, these terms are more generally applicable and have been adopted by IUPAC[2]. The prefixes *cis-* and *trans-* continue to be useful for cyclic compounds, but *E-* and *Z-* are taking over for oximes and the like as well as for olefinic linkages.

For the rather intricate details on how to combine all these symbols with names the IUPAC rules[2] should be consulted.

The conformational terms equatorial, axial, pseudoequatorial, pseudoaxial, chair, boat, twist, staggered, eclipsed, etc., are firmly established with their textbook meanings, and a variety of other terms such as tub and crown are familiar to specialists. Synperiplanar, synclinal, anticlinal and antiperiplanar, which define specific conformations around bonds, are also in wide use.

There are, however, areas where the sequence rule is not useful, namely where genetic relations between a large number of compounds have to be borne in mind. In such areas specialist nomenclature is adopted, e.g. D/L for carbohydrates, cyclitols, amino acids and peptides*; and α/β for steroid and higher terpenes*. Moreover, the sequence-rule procedure for octahedral complexes has not found favor with the IUPAC Commission for Nomenclature of Inorganic Chemistry, who have proposed a different approach[6], with a symbolism $\Delta\Lambda\delta\lambda$ that is controlled more by the geometry and less by the constitutional nature of the ligands.

Of the newer developments, the important terms prochiral and prochirality[7] are in general use, though possibly subject to slight revision in detailed application[8].

Much interest attaches to developments of topism (or topicity), which concerns the relation of groups or atoms to their environments. This leads to pairs of corresponding terms: homotopic/identical; heterotopic/isomeric, constitutionally heterotopic/constitutionally isomeric, stereoheterotopic/stereoisomeric, enantiotopic/enantiomeric, diastereotopic/diastereoisomeric. These concepts have been built by Hirschmann and Hanson[8,9] into a new approach to analysis of stereoisomerism and prostereoisomerism involving rigid definitions of ligand groups and centers of stereoisomerism and prostereoisomerism.

* See Chapter 7 for references.

A completely different approach by Prelog[10] is based on combinations of simplexes, considered two-dimensionally and three-dimensionally; an algebraic analysis is given by Ruch[11].

Whilst this book is not the place to set out these ideas it is perhaps the place to issue a warning that they may alter our whole approach to stereochemistry and thus also to its vocabulary and nomenclature.

REFERENCES

1. *IUPAC Nomenclature of Organic Chemistry, Definitive Rules for: Section A. Hydrocarbons; Section B. Fundamental Heterocyclic Systems; Section C. Characteristic Groups Containing Carbon, Hydrogen, Oxygen, Nitrogen, Halogen, Sulfur, Selenium and/or Tellurium.* 1969. A,B 3rd edn.; C 2nd edn., Butterworths, London (1971). (a) pp. 134–143; (b) pp. 186, 187, 190, 193, 194; (c) pp.261–268; (d) pp. 277–303
2. *IUPAC 1968 Tentative Rules for the Nomenclature of Organic Chemistry, Section E, Fundamental Stereochemistry, IUPAC Information Bulletin,* No. 35, 36–80 (June 1969); *J. Org. Chem.,* **35**, 2849 (1970)
3. Cahn, R. S., Ingold, Sir C. and Prelog, V., *Angew. Chem. Internat. Ed.,* **5**, 385, 511 (1966)
4. Cahn, R. S., *J. Chem. Educ.,* **41**, 116 (1964); also Ref. 2, Appendix 2
5. Blackwood, J. E., Gladys, C. L., Loening, K. L., Petrarca, A. E. and Rush, J. E., *J. Am. Chem. Soc.,* **90**, 509 (1968)
6. *IUPAC Nomenclature of Inorganic Chemistry,* 2nd edn., *Definitive Rules 1970,* Butterworths, London (1971) [reprinted from *Pure and Appl. Chem.,* **28**, 75–83 (1971)]
7. Hanson, K. R., *J. Am. Chem., Soc.,* **88**, 2731 (1966)
8. Hirschmann, H. and Hanson, K. R., *Eur. J. Biochem.,* **22**, 301 (1971)
9. Hirschmann, H. and Hanson, K. R., *J. Org. Chem.,* **36**,3 293 (1971); cf. Mislow, K. and Raban, M., *Topics in Stereochemistry,* **1**, 1 (1967)
10. Prelog, V., *Chemistry in Britain,* **4**, 382 (1972); Prelog, V. and Helmchen, G., *Helv. Chim. Acta,* **55**, 2581 (1972).
11. Ruch, E., *Accounts of Chemical Research,* **5**, 49 (1972)

7   Specialist Nomenclature

This chapter consists of notes on specialist nomenclature that has been the subject of authoritative pronouncements; readily available references are cited for study of the detail.

ORGANOMETALLIC COMPOUNDS

Section D of the IUPAC Nomenclature of Organic Chemistry is planned to deal with organic compounds of all the elements other than H, O, N, Halogen, S, Se and Te, these seven types having been treated in Section C. Various drafts of this further Section D have been discussed during some years jointly by the IUPAC Commissions on the Nomenclature of Inorganic and of Organic Chemistry; but as yet no recommendations have been published by IUPAC, even on a tentative basis*, apart from sections in the inorganic nomenclature rules that can be held to fall within the scope of organic Section D.

There still remains, therefore, the nomenclature for organophosphorus compounds that was agreed in 1952 by committees of the (British) Chemical Society and of the Organic Division of the American Chemical Society[1]. When originally published it brought order into conflicting practices and contained new methods of solving some old problems. Objections have been raised, almost wholly on the Continent of Europe, that some of the science on which the rules are based is outdated and that some of the names are too long or too novel. Nevertheless, these rules have been used for 20 years in *Chemical Abstracts* and by chemists of many, but not all, countries with satisfaction. It remains to be seen what improvements or alterations will be recommended by the IUPAC Commissions.

CARBOHYDRATES

In 1952, also, conflicting British and American practices were brought into line for carbohydrates. The resulting rules[2] were slightly modified some years later[3] and formed the basis of recent

---

* It is hoped that these Tentative Rules will be available in 1973 for purchase from Butterworths, 88 Kingsway, London WC2B 6AB, or from the IUPAC Secretariat.

IUPAC/IUB* rules[4]. However, these latest rules abandon one major change in classical carbohydrate chemistry that originated in USA but was at once adopted in the joint Anglo-American rules of 1952 as highly desirable. This change was that the configurational prefixes gluco, arabo, etc., should refer to consecutive but not necessarily contiguous asymmetric centers in the chain. Previously these centers had to be contiguous, i.e. not separated by CH_2 or CO groups, and the latest IUB/IUPAC rules revert to this practice. However, the IUB/IUPAC rules contain a 'note to Rule Carb-8' setting out, without comment, the previous Anglo-American usage; also, the Anglo-American names for individual examples are given when they differ from the IUPAC/IUB names (this occurs mainly with deoxy sugars and certain ketoses); the IUPAC/IUB objective appears to be to disapprove the Anglo-American procedure but to illustrate its correct use by those who insist on using it.

STEROIDS

This is the third large area where order was introduced during the 1950s. The rules, whose first version was published in 1957, have been amended several times by IUB/IUPAC discussions. The latest version, a second Definitive version, is published by IUPAC[5]. Several slightly earlier versions have been published in research journals[6], and reprints of these with amendments to bring them into line with the latest Definitive versions can be obtained from IUB*.

Each new version has extended the previous one, sometimes by additional new principles, as well as by correcting errors or clarifying drafting. The present rules are valid also for large areas of terpene chemistry and have found wide acceptance.

Among the important generally applicable features developed are the α/β nomenclature for stereochemistry around the tetracyclic nucleus but with adoption of the sequence rule for other centers of chirality (except C-20), together with regulation of nor and homo for ring contraction and expansion, seco for ring fission, and abeo for rearrangements.

* In this chapter, rules designated IUPAC/IUB are those issued jointly by the IUPAC Commission on the Nomenclature of Organic Chemistry and the IUPAC/IUB Commission on Biochemical Nomenclature. Reprints of these rules as published in research journals can be obtained gratis from the Office of Biochemical Nomenclature (Director, Dr. Waldo E. Cohn, Biology Division, Oak Ridge National Laboratory, P.O. Box Y, Oak Ridge, Tennessee 37830, USA; or as IUPAC publications (when available) from IUPAC Secretariat, Bank Court Chambers, 2/3 Pound Way, Cowley Centre, Oxford OX4 3YF, England.

POLYMERS

Tentative Rules for Nomenclature of Regular Single-Strand Organic Polymers have very recently been published by IUPAC as Appendix 29 to the IUPAC *Information Bulletin**.

PHYSICOCHEMICAL

Probably the most useful publication of its kind is the 1971 Report, 'Quantities, Units, and Symbols', from the Symbols Committee of the Royal Society†. This is compiled from the recommendations of the five international bodies concerned and includes only terms or symbols that have been internationally agreed. It includes, of course, the SI system of units; those wanting more detailed exposition of this can have recourse to one of the several books available‡.

Even the most experimental of non-physical chemists is affected by the changes introduced. He should note, for instance, that 'the mole is the amount of substance of a system which contains as many elementary units as there are atoms in 0.012 kilograms of carbon 12', so that there are now moles of atoms, ions, or electrons, etc., as well as of molecules; he may also see that 'mole' is the name of the unit, but 'mol' is its symbol (just as meter is a unit and m is its symbol); also that a liter is now reinstated as a special name for a cubic decimeter and no longer has a value of $1.000\ 028\ dm^3$.

We may also list here the prefixes to be used to construct decimal multiples of units:

Multiple	Prefix	Symbol	Multiple	Prefix	Symbol
10^{-1}	deci	d	10	deca	da
10^{-2}	centi	c	10^2	hecto	h
10^{-3}	milli	m	10^3	kilo	k
10^{-6}	micro	μ	10^6	mega	M
10^{-9}	nano	n	10^9	giga	G
10^{-12}	pico	p	10^{12}	tera	T
10^{-15}	femto	f			
10^{-18}	atto	a			

* Available from IUPAC Secretariat, Bank Court Chambers, 2/3 Pound Way, Cowley Centre, Oxford OX4 3YF, England; or from Butterworths, 88 Kingsway, London WC2B 6AB, England.

† Obtainable, £0·35 per copy, £6 for 25 copies (post free) from the Royal Society, 6 Carlton House Terrace, London, SW1Y 5AG, England.

‡ For example, Socrates, G. and Sapper, L. J., *SI and Metrication Conversion Tables*, Butterworths, London (1970).

Various recommendations on physical chemistry have been issued by IUPAC. The following Definitive versions are available for purchase*:

Manual of Definitions, Terminology and Symbols in Colloid and Surface Chemistry, Part I (£0·75). Part II is likely to be published in the IUPAC journal, *Pure and Applied Chemistry*, during 1974.
Presentation of NMR Data for Publication in Chemical Journals.

The following are among Tentative Recommendations now published as numbered Appendices to the IUPAC *Information Bulletin* and are likely to be made Definitive (perhaps after amendment) in 1973; they will then be due for publication in *Pure and Applied Chemistry*:

Nos. 1 and 27. Nomenclature, Symbols, Units and their Usage in Spectrochemical Analysis, Parts I, II, and III. (Professor C. Th. J. Alkemade has kindly informed me that the symbol M for atomic weight in Table 2.1 of Part III should be changed to A_r.
No. 5. Ion-Exchange Nomenclature.
No. 14. Nomenclature for Contamination Phenomena in Precipitation from Aqueous Solutions.
No. 15. Nomenclature for Chromatography.
No. 16. Nomenclature of Thermal Analysis.
No. 17. Nomenclature of Mass Spectrometry.
No. 18. Nomenclature of Scales of Working in Analysis.
No. 24. Names and Symbols for Light and Related Electromagnetic Radiations.
No. 25. Nomenclature for Nuclear Chemistry.
No. 28. Electrochemical Definitions and Symbols.

BIOCHEMICAL/CHEMICAL

The following recommendations from the IUPAC/IUB Commission on Biochemical Nomenclature will be of interest to chemists and biochemists working in the fields concerned; reprints are available†.

Abbreviations and Symbols for Chemical Names of Special Interest in Biological Chemistry: *Biochem. J.*, **101**, 1 (1966); *Biochemistry*, **5**, 1445 (1966).

Amino Acids (names): *J. Am. Chem. Soc.*, **82**, 5575 (1960); *J. Org. Chem.*, **28**, 291 (1963). Under revision.

* From IUPAC Secretariat, Bank Court Chambers, 2/3 Pound Way, Cowley Centre, Oxford OX4 3YF, England; or from Butterworths, 88 Kingsway, London WC2B 6AB, England.
† See footnote * on p. 119.

Amino Acids and Peptides (symbols): *Biochem. J.*, **126**, 773 (1972); *J. Biol. Chem.*, **249**, 977 (1972).

Peptides, Synthetic Modifications of: *Biochem. J.*, **104**, 17 (1967); *Biochemistry*, **6**, 362 (1967).

Peptides, Synthetic (Polymerized Amino Acids), Abbreviated Nomenclature: *Biochem. J.*, **127**, 753 (1972); *Biochemistry*, **11**, 942 (1972).

Peptide Chains, Conformations of: *Biochem. J.*, **121**, 577 (1971); *Biochemistry*, **9**, 3471 (1970).

Carotenoids: *Eur. J. Biochem.*, **25**, 397 (1972); *Biochemistry*, **10**, 4827 (1971).

Cyclitols: *Biochem. J.*, **112**, 17 (1969); *Biochim. Biophys. Acta*, **165**, 1 (1968).

Lipids: *Eur. J. Biochem.*, **2**, 127 (1967); and **12**, 1 (1970); *Biochim. Biophys. Acta*, **152**, 1 (1968); and **202**, 404 (1970).

Nucleic Acids, Polynucleotides, and Constituents: *Biochem. J.*, **120**, 449 (1971); *Biochemistry*, **9**, 4022 (1970).

Vitamins (and miscellaneous other compounds): *Biochem. J.*, **102**, 15 (1967); *J. Biol. Chem.*, **241**, 2987 (1966).

The following Tentative Recommendations, published as numbered Appendices to the IUPAC *Information Bulletin*, are likely to be included (possibly after amendment) in the IUPAC journal, *Pure and Applied Chemistry*, during 1974:

No. 20. Quantities and Units in Clinical Chemistry.

No. 21. List of Quantities in Clinical Chemistry.

REFERENCES

1. *J. Chem. Soc.*, 5122 (1952); *Chem. Eng. News*, **30**, 4515 (1952)
2. *J. Chem. Soc.*, 5108 (1952)
3. *J. Org. Chem.*, **28**, 281 (1963)
4. *Biochem. J.*, **125**, 673 (1971); *Biochemistry*, **10**, 3983 (1971)
5. *Pure and Appl. Chem.*, **31**, Nos. 1–2 (1972)
6. For example, *Biochemistry*, **8**, 2227 (1969); and **10**, 4994 (1971); *Biochem. J.*, **113**, 5 (1969); and **127**, 613 (1972)

Index